D0742075

PROGRESS IN CLINICAL AND BIOLOGICAL RESEARCH

RECENT TITLES

Vol 163: **Prevention of Physical and Mental Congenital Defects**, Maurice Marois, *Editor*. Published in 3 volumes: Part A: *The Scope of the Problem*. Part B: *Epidemiology, Early Detection and Therapy, and Environmental Factors*. Part C: *Basic and Medical Science, Education, and Future Strategies*

Vol 164: **Information and Energy Transduction in Biological Membranes,** C. Liana Bolis, Ernst J.M. Helmreich, Hermann Passow, *Editors*

Vol 165: **The Red Cell: Sixth Ann Arbor Conference**, George J. Brewer, *Editor*

Vol 166: **Human Alkaline Phosphatases,** Torgny Stigbrand, William H. Fishman, *Editors*

Vol 167: **Ethopharmacological Aggression Research**, Klaus A. Miczek, Menno R. Kruk, Berend Olivier, *Editors*

Vol 168: **Epithelial Calcium and Phosphate Transport: Molecular and Cellular Aspects,** Felix Bronner, Meinrad Peterlik, *Editors*

Vol 169: **Biological Perspectives on Aggression,** Kevin J. Flannelly, Robert J. Blanchard, D. Caroline Blanchard, *Editors*

Vol 170: **Porphyrin Localization and Treatment of Tumors,** Daniel R. Doiron, Charles J. Gomer, *Editors*

Vol 171: **Developmental Mechanisms: Normal and Abnormal,** James W. Lash, Lauri Saxén, *Editors*

Vol 172: **Molecular Basis of Cancer,** Robert Rein, *Editor*. Published in 2 volumes: Part A: *Macromolecular Structure, Carcinogens, and Oncogenes*. Part B: *Macromolecular Recognition, Chemotherapy, and Immunology*

Vol 173: **Pepsinogens in Man: Clinical and Genetic Advances,** Johanna Kreuning, I. Michael Samloff, Jerome I. Rotter, Aldur Eriksson, *Editors*

Vol 174: **Enzymology of Carbonyl Metabolism 2: Aldehyde Dehydrogenase, Aldo-Keto Reductase, and Alcohol Dehydrogenase,** T. Geoffrey Flynn, Henry Weiner, *Editors*

Vol 175: **Advances in Neuroblastoma Research,** Audrey E. Evans, Giulio J. D'Angio, Robert C. Seeger, *Editors*

Vol 176: **Contemporary Sensory Neurobiology,** Manning J. Correia, Adrian A. Perachio, *Editors*

Vol 177: **Medical Genetics: Past, Present, Future,** Kåre Berg, *Editor*

Vol 178: **Bluetongue and Related Orbiviruses,** T. Lynwood Barber, Michael M. Jochim, *Editors*

Vol 179: **Taurine: Biological Actions and Clinical Perspectives,** Simo S. Oja, Liisa Ahtee, Pirjo Kontro, Matti K. Paasonen, *Editors*

Vol 180: **Intracellular Protein Catabolism,** Edward A. Khairallah, Judith S. Bond, John W.C. Bird, *Editors*

Vol 181: **Germfree Research: Microflora Control and Its Application to the Biomedical Sciences,** Bernard S. Wostmann, *Editor*, Julian R. Pleasants, Morris Pollard, Bernard A. Teah, Morris Wagner, *Co-Editors*

Vol 182: **Infection, Immunity, and Blood Transfusion,** Roger Y. Dodd and Lewellys F. Barker, *Editors*

Vol 183: **Aldehyde Adducts in Alcoholism,** Michael A. Collins, *Editor*

Vol 184: **Hematopoietic Stem Cell Physiology**, Eugene P. Cronkite, Nicholas Dainiak, Ronald P. McCaffrey, Jiri Palek, Peter J. Quesenberry, *Editors*

Vol 185: **EORTC Genitourinary Group Monograph 2,** Fritz H. Schroeder, Brian Richards, *Editors*. Published in two volumes: Part A: *Therapeutic Principles in Metastatic Prostatic Cancer*. Part B: *Superficial Bladder Tumors*

Vol 186: **Carcinoma of the Large Bowel and Its Precursors,** John R.F. Ingall, Anthony J. Mastromarino, *Editors*

Vol 187: **Normal and Abnormal Bone Growth: Basic and Clinical Research,** Andrew D. Dixon, Bernard G. Sarnat, *Editors*

Vol 188: **Detection and Treatment of Lipid and Lipoprotein Disorders of Childhood,** Kurt Widhalm, Herbert K. Naito, *Editors*

Vol 189: **Bacterial Endotoxins: Structure, Biomedical Significance, and Detection With the Limulus Amebocyte Lysate Test,** Jan W. ten Cate, Harry R. Büller, Augueste Sturk, Jack Levin, *Editors*

Vol 190: **The Interphotoreceptor Matrix in Health and Disease,** C. David Bridges, *Editor*, Alice J. Adler, *Associate Editor*

Vol 191: **Experimental Approaches for the Study of Hemoglobin Switching,** George Stamatoyannopoulos, Arthur W. Nienhuis, *Editors*

Vol 192: **Endocoids,** Harbans Lal, Frank LaBella, John Lane, *Editors*

Vol 193: **Fetal Liver Transplantation,** Robert Peter Gale, Jean-Louis Touraine, Guido Lucarelli, *Editors*

Vol 194: **Diseases of Complex Etiology in Small Populations: Ethnic Differences and Research Approaches,** Ranajit Chakraborty, Emöke J.E. Szathmary, *Editors*

Vol 195: **Cellular and Molecular Aspects of Aging: The Red Cell as a Model,** John W. Eaton, Diane K. Konzen, James G. White, *Editors*

Vol 196: **Advances in Microscopy,** Ronald R. Cowden, Fredrick W. Harrison, *Editors*

Vol 197: **Cooperative Approaches to Research and Development of Orphan Drugs,** Melvin Van Woert, Eunyong Chung, *Editors*

Vol 198: **Biochemistry and Biology of DNA Methylation,** Giulio L. Cantoni, Aharon Razin, *Editors*

Vol 199: **Leukotrienes in Cardiovascular and Pulmonary Function,** Allan M. Lefer, Marlys H. Gee, *Editors*

Vol 200: **Endocrine Genetics and Genetics of Growth,** Costas J. Papadatos, Christos S. Bartsocas, *Editors*

Vol 201: **Primary Chemotherapy in Cancer Medicine,** D.J. Theo Wagener, Geert H. Blijham, Jan B.E. Smeets, Jacques A. Wils, *Editors*

Vol 202: **The 2-5A System: Molecular and Clinical Aspects of the Interferon-Regulated Pathway,** Bryan R.G. Williams, Robert H. Silverman, *Editors*

Vol 203: **Testicular Cancer,** Saad Khoury, René Küss, Gerald P. Murphy, Christian Chatelain, James P. Karr, *Editors*

Vol 204: **Tumor Markers and Their Significance in the Management of Breast Cancer,** Thomas Dao, Angela Brodie, Clement Ip, *Editors*

Vol 205: **Comparative Endocrinology: Developments and Directions,** Charles L. Ralph, *Editor*

Vol 206: **Genetic Toxicology of the Diet,** Ib Knudsen, *Editor*

Vol 207: **Monitoring of Occupational Genotoxicants,** Marja Sorsa, Hannu Norppa, *Editors*

Vol 208: **Risk and Reason: Risk Assessment in Relation to Environmental Mutagens and Carcinogens,** Per Oftedal, Anton Brøgger, *Editors*

Vol 209: **Genetic Toxicology of Environmental Chemicals,** Claes Ramel, Bo Lambert, Jan Magnusson, *Editors*. Published in two volumes: Part A: *Basic Principles and Mechanisms of Action.* Part B: *Genetic Effects and Applied Mutagenesis*

Vol 210: **Ionic Currents in Development,** Richard Nuccitelli, *Editor*

Vol 211: **Transfusion Medicine: Recent Technological Advances,** Kris Murawski, Frans Peetoom, *Editors*

Vol 212: **Cancer Metastasis: Experimental and Clinical Strategies,** D.R. Welch, B.K. Bhuyan, L.A. Liotta, *Editors*

Vol 213: **Plant Flavonoids in Biology and Medicine: Biochemical, Pharmacological, and Structure–Activity Relationships,** Vivian Cody, Elliott Middleton, Jr., Jeffrey B. Harborne, *Editors*

Vol 214: **Ethnic Differences in Reactions to Drugs and Xenobiotics,** Werner Kalow, H. Werner Goedde, Dharam P. Agarwal, *Editors*

Vol 215: **Megakaryocyte Development and Function,** Richard F. Levine, Neil Williams, Jack Levin, Bruce L. Evatt, *Editors*

Vol 216: **Advances in Cancer Control: Health Care Financing and Research,** Lee E. Mortenson, Paul F. Engstrom, Paul N. Anderson, *Editors*

Vol 217: **Progress in Developmental Biology,** Harold C. Slavkin, *Editor.* Published in two volumes.

Vol 218: **Evolutionary Perspective and the New Genetics**, Henry Gershowitz, Donald L. Rucknagel, Richard E. Tashian, *Editors*

Vol 219: **Recent Advances in Arterial Diseases: Atherosclerosis, Hypertension, and Vasospasm,** Thomas N. Tulenko, Robert H. Cox, *Editors*

Vol 220: **Safety and Health Aspects of Organic Solvents,** Vesa Riihimäki, Ulf Ulfvarson, *Editors*

Please contact the publisher for information about previous titles in this series.

Safety and Health Aspects of Organic Solvents

Safety and Health Aspects of Organic Solvents

Proceedings of the International Course on Safety and Health Aspects of Organic Solvents Held in Espoo, Finland, April 22-26, 1985

Editors

Vesa Riihimäki
Institute of Occupational Health
Department of Industrial Hygiene
and Toxicology
Helsinki, Finland

Ulf Ulfvarson
Department of Work Science
The Royal Institute of Technology
Stockholm, Sweden

ALAN R. LISS, INC. • NEW YORK

Address all Inquiries to the Publisher
Alan R. Liss, Inc., 41 East 11th Street, New York, NY 10003

Printed in the United States of America

Library of Congress Cataloging in Publication Data

International Course on Safety and Health Aspects of
Organic Solvents (1985 : Espoo, Finland)
Safety and health aspects of organic solvents.

(Progress in clinical and biological research ; v. 220)
Includes bibliographies and index.
1. Nonaqueous solvents—Toxicology—Congresses.
2. Solvents industry—Hygienic aspects—Congresses.
3. Nonaqueous solvents—Safety measures—Congresses.
I. Riihimäki, Vesa. II. Ulfvarson, Ulf, 1931–
III. Title. IV. Series. [DNLM: 1. Occupational
Diseases—chemically induced—congresses. 2. Solvents—
adverse effects—congresses. W1 PR668E v.220 /
QV 633 I595s 1985]
RA 1270.S6I58 1985 615.9′02 86-20137
ISBN 0-8451-5070-7

Contents

Contributors

Klaus E. Andersen, Dermatology Clinic, Roskilde, Denmark **[133]**

Mari Antti-Poika, Institute of Occupational Health, Helsinki, Finland **[255]**

Alf Askergren, Construction Industry's Organisation for Working Environment, Safety and Health, Stockholm, Sweden **[155]**

Karl-Heinz Cohr, Department of Chemistry and Toxicology, National Institute of Occupational Health, Hellerup, Denmark **[45]**

Martin Døssing, Medical Department F, Gentofte University Hospital, Hellerup, Denmark **[97]**

Pierre O. Droz, Institute of Occupational Medicine and Industrial Hygiene, University of Lausanne, LeMont, Switzerland **[73]**

Francesco Gamberale, Research Unit of Psychophysiology, Occupational Health Department, National Board of Occupational Safety and Health, Solna, Sweden **[203]**

J. Gmehling, Universität Dortmund, Abteilung Chemietechnik, Dortmund, Federal Republic of Germany **[31]**

Paul Grasso, Robens Institute, University of Surrey, Guildford, Surrey, UK **[187]**

Helena Hänninen, Department of Psychology, Institute of Occupational Health, Helsinki, Finland **[225]**

Charles M. Hansen, Scandinavian Paint and Printing Ink Research Institute, Horsholm, Denmark; present address: Jens Bornøsvej 16, Hørsholm, Denmark **[297]**

Peter C. Holmberg, Department of Epidemiology and Biostatistics, Institute of Occupational Health, Helsinki, Finland **[179]**

Juhani Jaakkola, Department of Industrial Hygiene and Toxicology, Institute of Occupational Health, Helsinki, Finland **[303]**

Juhani Juntunen, Clinical Neurosciences, Institute of Occupational Health, Helsinki, Finland **[265]**

Pentti Kalliokoski, Department of Industrial Hygiene, University of Kuopio, Kuopio, Finland **[21]**

Matti Klockars, Institute of Occupational Health, Helsinki, Finland **[139]**

The number in brackets is the opening page number of the contributor's article.

Eeva Kuosma, Department of Epidemiology and Biostatistics, Institute of Occupational Health, Helsinki, Finland **[179]**

Kari Kurppa, Department of Epidemiology and Biostatistics, Institute of Occupational Health, Helsinki, Finland **[179]**

Arto Laine, Department of Industrial Hygiene and Toxicology, Institute of Occupational Health, Helsinki, Finland **[123]**

E. Lehmann, Bundesanstalt für Arbeitsschutz, Dortmund, Federal Republic of Germany **[31]**

Juha Nickels, Department of Occupational Medicine, Institute of Occupational Health, Helsinki, Finland **[115]**

M.G. Parkki, Department of Physiology, University of Turku, Turku, Finland **[89]**

Olavi Pelkonen, Department of Pharmacology, University of Oulu, Oulu, Finland **[107]**

Kaarina Rantala, Uusimaa Regional Institute of Occupational Health, Helsinki, Finland **[179]**

Riitta Riala, Uusimaa Regional Institute of Occupational Health, Helsinki, Finland **[179]**

Vesa Riihimäki, Department of Industrial Hygiene and Toxicology, Institute of Occupational Health, Helsinki, Finland **[xi,61,123]**

Anna Maria Seppäläinen, Division of Clinical Neurophysiology, Department of Neurology, University of Helsinki, Helsinki, Finland **[247]**

Ole Svane, Danish Labour Inspection Service, Copenhagen, Denmark **[283]**

Ulf Ulfvarson, Research Department, Swedish National Board of Occupational Safety and Health, Stockholm, Sweden; present address: Department of Work Science, The Royal Institute of Technology, Stockholm, Sweden **[xi,3]**

Gyorgy Ungváry, National Institute of Occupational Health, Budapest, Hungary **[169]**

U. Weidlich, Universität Dortmund, Abteilung Chemietechnik, Dortmund, Federal Republic of Germany **[31]**

Arne Wennberg, Department of Occupational Neuromedicine, Research Division, National Board of Occupational Safety and Health, Solna, Sweden **[237]**

Preface

Organic solvents remain an important group of occupational and domestic toxicants despite continuing efforts to replace them in a variety of products such as paints, varnishes, and adhesives. In the past, the main hazards of solvents were generally related to their flammability and acutely narcotizing properties. More recently the emphasis has shifted from accidental situations to hazards posed by low-level exposures over a long time or over an especially sensitive period of life. In the meantime modern industrial processes and advances in the techniques of industrial hygiene have helped to reduce the exposures of workers considerably. Yet the many manual trades which universally involve solvent exposures via inhalation and percutaneously are still a cause for concern.

Chronic poisoning by industrial solvents encompassing deleterious effects on the nervous system has been a controversial issue in Scandinavia and elsewhere. While most researchers agree that such an affection exists, indeed the central nervous system appears to be the primary target of solvent toxicity in man, opinion has been divided on the types and severity of effects involved. The past experience has been fraught with problems of recognizing the effects by solvents from other clinical neuropsychological and neurological states, and many epidemiological studies have been flawed in design to give any indication of causality. Thus when establishing a diagnosis of solvent poisoning, a certain set of criteria needs to be fulfilled, a procedure recently endorsed by a joint WHO/Nordic Council of Ministers Working Group.

The mechanisms and pathogenesis of solvent toxicity in humans are still very much open questions; factors of individual susceptibility, compensating capacity, and toxic interactions arising in the typically multicomponent exposures may play a large role. Moreover, effects, other than the nervous ones such as nephrotoxicity, carcinogenicity, reproductive effects, and teratogenicity have become new objects of investigation.

This volume constitutes the proceedings of the International Course on Safety and Health Aspects of Organic Solvents, held in Espoo, Finland, April 22-26, 1985. The financial support by the Nordic Council of Ministers is hereby gratefully acknowledged.

The proceedings presents an overview of the use and occurrence of organic solvents in the work environment of industrialized countries, the factors governing uptake and tissue distribution of solvents, events leading to toxicity such as metabolism, occupational toxicology, and prevention of solvent hazards. The book is aimed for all professional groups concerned with the safety and health aspects of solvents and it should find particular use for the motivation and guidance of the occupational health personnel.

Vesa Riihimäki
Ulf Ulfvarson

UTILIZATION AND OCCURRENCE

Safety and Health Aspects of Organic Solvents, pages 3–19

ORGANIC SOLVENTS: CONCEPT, UTILIZATION IN INDUSTRY AND OCCUPATIONAL EXPOSURE

Ulf Ulfvarson
Swedish National Board of Occupational Safety and Health, Stockholm

INTRODUCTION

Solutions have an importance in nature, sciences and technology which can hardly be overestimated. The solvents are important as media for chemical reactions and as means of material transportation on the molecular as well as on the macroscopic level. The importance of water as a reaction medium and as a means of transportation in the streams of nature and in living organisms needs no elaboration. There are other inorganic solvents besides water that are of importance in technology, e.g. carbon disulphide, but while the number of inorganic solvents is rather limited there is a profusion of organic substances used as solvents.

CHARACTERISTIC PROPERTIES OF ORGANIC SOLVENTS

Organic substances suitable as solvents may be characterized in the following way. They must be relatively inexpensive to produce. At atmospheric pressure they have a boiling point not much lower than 0 °C or higher than 200 °C. Ordinarily they have a small or moderate chemical reactivity towards most common materials.

Although there may be many exceptions to this description, it will allow a useful discussion of organic solvents as a group and of their uses.

The solvent function is not necessarily the sole use of the organic substances in question, nor even the most important one. Methanol for instance is used mostly as a chemical in the organic synthesis of formaldehyde.

The organic solvents have to be relatively inexpensive, since in many technical applications they are not reused but are supposed to evaporate to the environment after fulfilling their purpose.

Although most organic solvents are liquid at room temperature, there is the possibility of using easily condensable gases under pressure as solvents. An example is the application of a mixture of propane and butane under pressure in household aerosols to dissolve such things as insecticides.

The boiling point must not be too high, and consequently the vapour pressure too low, since otherwise the solvents may be difficult to separate from the solutes or from the cleaned goods in cleaning operations.

The organic solvents ought to have a small or at the most a moderate chemical reactivity. If they are too reactive, there is the risk of chemical reactions with the solutes or the material in equipment or containers. This is a rule with exceptions however, since in some compositions, e.g. glues and lacquers, the solvents are monomers which will partly polymerize in the curing process, as is the case for example with styrene in polystyrene.

The general lack of chemical reactivity amongst the organic solvents is also a prerequisite for their use and abuse in technology. It would be hardly conceivable to allow particularly reactive substances to evaporate more or less unimpededly to the environment. Therefore even if organic acids, amines or aldehydes could serve a purpose as solvents in special applications they would irritate mucuous membranes and might also be corrosive to the materials in cans, etc. Quite another thing is that the lack of reactivity is often illusory, from the biochemical point of view. Substances which were formerly assumed to be excreted unchanged or as harmless reaction products from the human organism have in due course been shown to be harmful to the organs of the body and have therefore been abandoned as technical solvents. Examples are benzene, carbon tetrachloride and methylbutylketone.

KINDS OF SUBSTANCES USED AS ORGANIC SOLVENTS

Common examples of organic substances used as solvents in various applications are aliphatic hydrocarbons from propane to decane, terpenes, aromatic hydrocarbons, mostly toluene and the xylenes, chlorinated hydrocarbons, primarily methylene chloride, trichloroethylene, tetrachloroethylene (or perchloroethylene), 1,1,1-trichloroethane (or methylchloroform), alcohols from methanol to butanol with isomers and dialcohols, mainly ethylene and propylene glycols and their ethers with methanol to butanol, the monoesters of the dialcohols and their ethers, ketones from acetone to methylisobutylketone and finally esters of acetic acid and a few lower alcohols.

When it comes to the kind of organic solvents used for various purposes it is necessary to consider a long-term trend in the different industries. The paint manufacturing industry may be taken as an example since it is a big direct consumer of solvents and since its products cause an important exposure to organic solvents in many other industries and occupations.

In 1920 most lacquers and varnishes were based on linseed oil, a binder, which has a low viscosity before drying and needs but a small addition of organic solvents to modify the viscosity. The solvent mostly used to this end was turpentine. The introduction of nitrocellulose lacquers in 1920 involved the use of alcohols, toluene and quite a lot of benzene. 1928 was the year of the birth of the alkyd resins and consequently the introduction of white spirit to be used as a solvent in lacquers and varnishes. The alkyd resins came into general use after the second world war. A number of new binders were introduced from 1930 onwards and consequently new organic solvents were called for. From 1932 to 1952 there was an almost ten fold increase in the procution of organic solvents in the USA, from 400 million tonnes to 3,000 million tonnes (Nylén and Sunderland, 1965). The most common organic solvents nowadays are toluene and the xylenes, butanol, sundry esters and ketones and methylene chloride. Not until the seventies have health and economic considerations to some extent hampered the use of organic solvents, when they were partly substituted for water. It is true that in some techniques solvents have been disposed of entirely as in powder paints, but in the broad picture these are unimportant exceptions. Organic solvents are still important in products for industrial painting and in printing inks. In glues the development towards water-based systems had advanced more rapidly, but organic solvents still exist for example in contact glues.

CONSUMPTION OF ORGANIC SOLVENTS

Let us now look at the consumption of organic solvents in a quantitative way. Industry in Sweden for instance annually consumes some hundred thousand tonnes of organic solvents. The figure could be made more exact if the definition of organic solvents were restricted to only those quantities which are used technically as solvents.

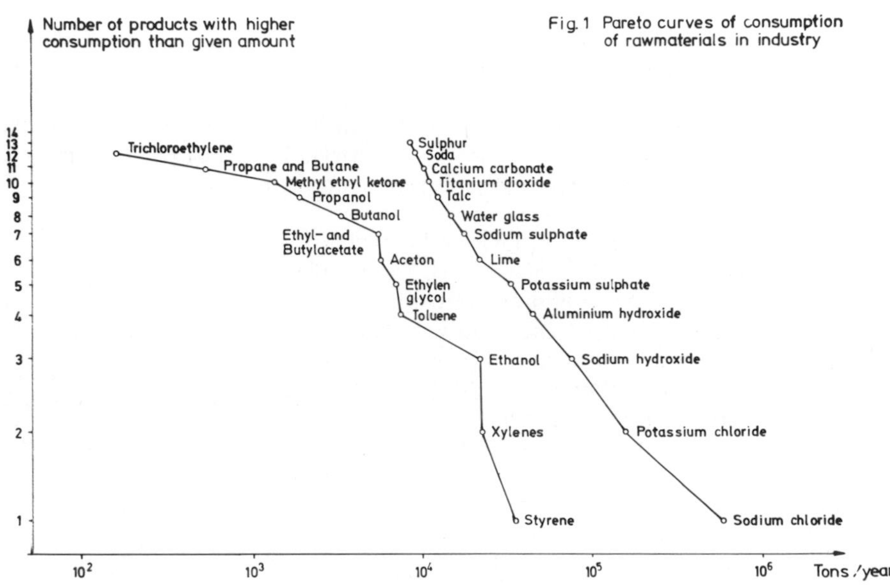

Fig. 1 Pareto curves of consumption of rawmaterials in industry

In Figure 1 the consumption in 1982 (Anon., 1984) of the most important organic substances used technically as organic solvents is plotted as a Pareto curve. As a comparison, the consumption of the most important inorganic bulk chemicals are plotted. Pareto used curves of this kind in order to analyse income differences between individuals in different countries (Anon., 1973-74). In a similar way the consumption differences are analysed here. When the slope of the curve is steep, the differences in the consumption of different substances are small and vice versa. It is reasonable to assume that a considerable difference in the consumption of two organic solvents is due to higher versatility in the solvent with the higher consumption. Many of the inorganic bulk chemicals are used for several purposes, and their Pareto curve is steep. The most important six or seven organic solvents can be used for many purposes or are interchangeable, so the differences in their consumption are small. Then comes a number of organic solvents with more specialized uses. Their properties or prices make them more or less unique and they can be exchanged only to a limited extent. Consequently they are characterized as a group by big differences in consumption, which leads to a flattening of the curve. An important conclusion from this analysis is that the substitution of a less hazardous substance for a substance which is found to be hazardous to health may be a troublesome process, taking a lot of development effort and time. This conclusion is also confirmed by experience, e.g. of the obvious difficulties of finding a substitute for trichloroethylene.

The emisson of organic solvents to the environment as air contamination is calculated as being about 125,000 tonnes annually in Sweden (Johansson et al., 1984). A considerable part comes from surface treatment, including degreasing operations, and the rest from a number of sources, for instance dry cleaning, chemical industry, storage, and loading and unloading of organic solvents. From the work-environment aspect it is natural to disregard some big emission sources since the resulting air pollution will not contaminate the work environment. This is the case in refineries and in petrochemical and organic plants, where the main emissons come from process ventilation. Even if unpurified, these emissons will ordinarily affect the employees only indirectly as general pollution.

The application of paints causes an emission of altogether 22,000 tonnes, dominated by the xylenes, white spirit, butyl acetate and butanol. The next biggest source of emission is printing, 7,300 tonnes of toluene, ethanol and aliphatic hydrocarbons, then comes plastic treatment with 3,300 tonnes of freons and styrene, engineering workshops with 3,000 tonnes (probably trichloroethylene and white spirit) and dry-cleaning shops with 1,500 tonnes (mainly tetrachloroethylene). The manufacturing of paint is a comparatively small source of emission of organic solvents, altogether 150 tonnes, which means that less than 1% of the total amount of organic solvents handled by the paint manufacturers is lost in the process.

The emisson of organic solvents from the food industry and the pharmaceutical industry is dominated by ethanol.

OCCUPATIONAL EXPOSURES - PREREQUISITES FOR EXPOSURE

We will now consider the possible and actual exposure to organic solvents in work operations and the supervision of production processes. To facilitate this discussion a few basic concepts have to be introduced.

Consider an operation causing an emission $\dot{m}$(kg/h) of an air pollutant in a workroom and a ventilation in this room consisting of two parts, a local exhaust ventilation with the capture efficiency α (between 0 and 1) and a general ventilation q (m^3/h). Then when equilibrium is reached and if no vortices are formed in the workroom, the concentration c (mg/m^3) in the locality will be

$$c = \frac{\dot{m}(1-\alpha)10^6}{q} \qquad \text{.... 1}$$

In order to calculate the equilibrium concentration we need a measure of the emission.

In industries consuming products with organic solvents which are supposed to evaporate entirely in the process the emission can be expressed by

$$\dot{m} = \frac{L}{N \cdot H} \qquad \text{.... 2}$$

L being the consumption (or loss) of organic solvents (kg/year), N the number of employees in contact with the relevant production and H the number of working hours per annum.

The hygienic effect E of the concentration is defined as

$$E = \frac{c}{C} \qquad \text{.... 3}$$

where C is the limit value. If several substances are present as air pollutants simultaneously, then

$$E = \sum_{i=1}^{n} c_i/C_i \qquad \text{.... 3a}$$

If equations 1, 2 and 3 are combined we get the average exposure in the workroom during working hours.

$$E = \frac{L \cdot (1-\alpha) \cdot 10^6}{N \cdot H \cdot q \cdot C} \quad \ldots. \; 4$$

We may estimate the demand on ventilation in different occupations if we assume that no local exhaust ventilation is employed, i.e. $\alpha = 0$, and that the general ventilation is according to minimum standards for workrooms without air pollution, i.e. 30 m^3 per h per employee (Olander, 1982). If the floor surface available for each employee is 3 m^2, this means 10 m^3 per h per m^2.

In Table 1 the exposures in a number of occupations are calculated given $q = 30\ m^3$ per h per employee; $\alpha = 0$. These exposures will of course rarely be found in practice, but the figures indicate the air pollution problem encountered in various trades. The most demanding problem is connected with painting, $E \leq 100$. Next comes dry cleaning $E \leq 33$ and printing offices and paint industries E 10-20. Even if the general ventilation were increased to the practical limit, say 10 to 20 times, the problem would not be solved. Obviously local exhaust ventilation will be necessary and we see from the measurements carried out in furniture factories, Table 1, that the control measures were successful. In these workrooms the painting was done by spraying in a ventilated box.

A second approach to an estimate of the risk of exposure is to calculate the emission of organic solvents from various sources like painted surfaces or open vessels containing solvents. There are a number of more or less empirical relations between rate of vaporization of an organic solvent and other parameters. In principle they are based on Fick's law of diffusion with corrections for convection. A useful formula is the following (Carson, 1976)

$$\dot{m} = 3.6 \cdot 10^{-10} \, (Mp^{o}/T) \, u^{0.76} r^{1.89} \quad \ldots. \; 5$$

In this formula $\dot{m}$ is the emission rate in g/s, M is the molecular weight of the organic solvent, p^o is the vapour pressure in dynes/cm^2, T is the absolute temperature, u is the air velocity in cm/s over an open vessel of radius r in cm. Combining equations 1, 3 and 5 gives a relationship between the width of an open vessel containing an organic solvent and the exposure in a room with a certain ventilation. In the resulting equation the quantity $\frac{p^o}{C}$ is included. This quantity may be expressed as the hygienic effect at saturation E_s of an organic solvent

Table 1. Demand on the ventilation in various occupations with exposure to organic solvents. The hygienic effect, E, calculated from equation 4 assuming $q = 30\ m^3$ per h per employee and $\alpha = 0$. Limit values according to ASF 1984:5. Production figures from ref 2 and 4, valid for 1982 in Swedish industry.

Occupation/ operation	1000 working hours	Emission, tonnes per annum								E	
		ethanol	perchloro-ethylene	toluene	xylene	white spirit	butanol	butyl acetate	others	cal-culated	observed
Dry cleaning	8740	-	1200	-	-	-	-	-	-	33	1-4
Paint manu-facturing	1910	-	-	10	40	40	-	-	60	17[a)]	10-100[b)]
Printing	33800	4000	-	1500	-	1100	-	-	700	15[a)]	2-3
Lacquering	53300			1000	6000	4000	1500	1500	6000	78[a)]	3[c)]
Lacquering in 3 furniture factories											
1	185	10.9	-	7.2	3.1	-	1.6	8.8	3.9	14	0.48
2	30	2.7	-	0.99	6.8	-	4.0	7.5	3.8	103	0.26
3	21	6.4	-	0.19	0.26	-	0.04	11.0	2.2	39	0.50

a) It is assumed that the hygienic limit value of "other" solvents is 25 ppm, which will probably lead to an overestimation of the risk

b) In manual cleaning of bulky pieces of equipment

c) In painting with a roller in a room with moderate ventilation

$$E_s = \frac{p^o \cdot 10^6 \cdot M}{760 \cdot C \cdot RT} \qquad \ldots. \; 6$$

where p^o is the vapour pressure in torrs and C as before the hygienic limit in mg/m^3. R is the gas constant 0.083 (l x bars/mol x ^{o}K). Combining the equations 1, 3, 5 and 6 and assuming the air velocity over the open vessel to be 0.5 m/sec and the general ventilation in the room to be 30 m^3 per h per employee gives us a formulation of the hygienic limit value in an unconventional, yet very concrete way, as the maximum diameter D of an open vessel kept filled to the brim with an organic solvent. At equilibrium the exposure caused by the evaporation from this vessel will be only just in compliance with the occupational health standard.

$$D = 2(14100/E_s)^{1/1.89} \qquad \ldots. \; 7$$

In Table 2, calculated maximum diameters are presented for a few organic solvents. The figures should be checked by experiment, but they are given for the sake of comparison of the risks of exposure to various solvents. Thus we see the following rule of thumb: The maximum size of the vessel corresponds to a barrel full of ethylene glycol, a bucket full of ethanol, a saucer full of acetone, a coffee cup full of trichloroethylene, an egg cup full of methylene chloride and a thimble full of benzene.

TABLE 2. Calculated maximum diameters D (cm) of open vessels, which if kept filled to the brim with organic solvents will cause air pollution only just in compliance with hygienic limit values when there is only a general ventilation of 30 m^3 per h per employee.

Substance	Hygienic effect at saturation E_s (equation 6)	D(cm)	Typical vessel
Ethylene glycol	<26	>56	barrel
Ethanol	66	34	bucket
Acetone	396	13	saucer
Trichloroethylene	2030	6	coffee cup
Methylene chloride	5800	3	egg cup
Benzene	12600	2	thimble

From these calculations it may be possible to get a rough idea of how various organic solvents may be handled. It may for instance be possible to wash windows with a rag dipped in ethanol without danger of exceeding the limit value for this solvent, while the continuous use of a spot remover containing trichloroethylene will cause non-compliance with the limit value if no control measures are taken. Experience from measurements supports this statement.

THE DIFFERENCE BETWEEN HANDLING AND CONSUMPTION

In the paint industry the intention is of course to keep as much of the organic solvents as possible in the products. The total loss of organic solvents in paint production has already been stated to be less than 1% of the quantity handled. The more volatile solvents will probably be lost in a higher proportion than the less volatile solvents. The exposure in the paint industry may therefore be calculated by derivations of the kind exemplified in equation 5.

The total exposure is also proportional to the production (=use) of each organic solvent. Thus if everything except the production of each organic solvent is kept constant the exposure will be

$$E \sim \sum_{i=1}^{n} P_i \cdot (E_s)_i \qquad \ldots . \; 8$$

P_i is the quantity of each organic solvent contained in the total production for sale.

In industries consuming products containing organic solvents all solvents will evaporate sooner or later. If the final evaporation of the organic solvents is brought about in a separate space, for instance in drying oven or a ventilated drying room, the exposure will be governed by equation 8 as in the paint industry, i.e. the different organic solvents will tend to pollute the working environment in proportion to their vapour pressures. If on the other hand the organic solvents evaporate in the work room as in the use of fast-drying paints and printing ink and in dry cleaning, the exposure will be governed by a combination of equations 1, 2 and 3, since even the organic solvents with low vapour pressure will evaporate in the work operation. To the extent determined by the local exhaust ventilation, they will reach the breathing zone of the worker.

The exposure will then be proportional to the loss L_i of each organic solvent and inversely proportional to the hygienic limit value C_i

$$E \sim \sum_{i=1}^{n} \frac{L_i}{C_i} \qquad \ldots . \ 9$$

The "consumption" of organic solvents in the paint industry during 1972-1982 is plotted in Figure 2 and the risk estimates according to equations 8 and 9 are plotted in Figure 3, as percentages of the respective average risk estimate during the period. The tendency is for the risk to slowly decrease over the period as evaluated by any indicator, owing to the decrease in the use of the rather volatile toluene and the general though gradual decrease in the total consumption of organic solvents. The tendency is somewhat more obvious in the paint industry (equation 8) than among the paint consumers (equation 9). In this hypothetical reasoning the change in manning and technology in the various industries has been purposely neglected.

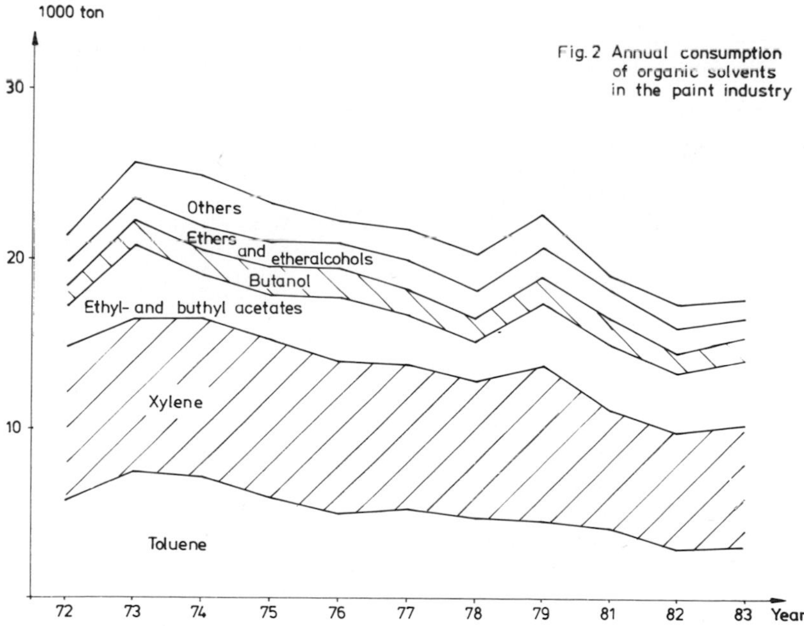

Fig. 2 Annual consumption of organic solvents in the paint industry

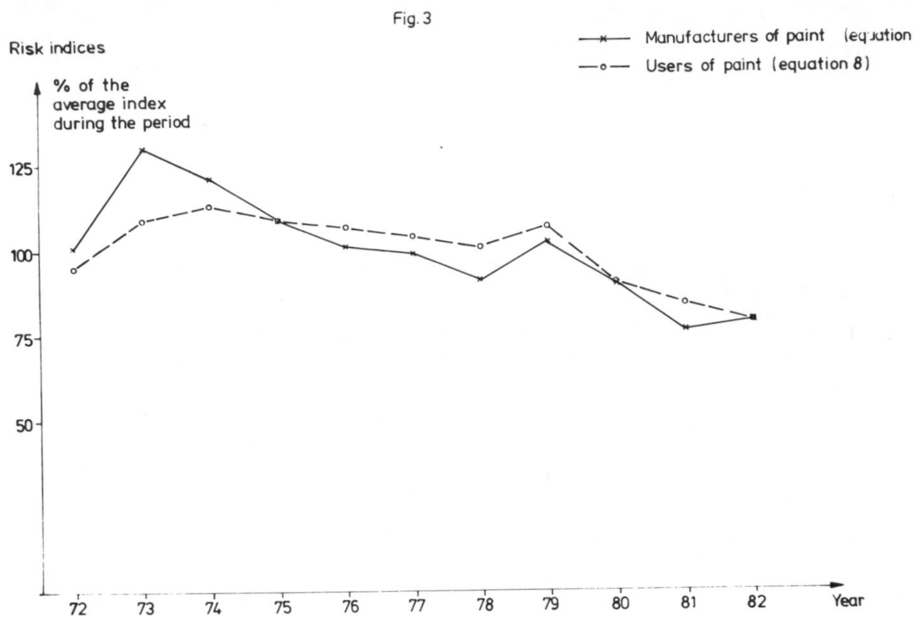

OBSERVED EXPOSURES, EXAMPLES

Generally speaking non-compliance with hygienic limit values is nowadays rare in Sweden, but in certain difficult operations excessive exposure is still found.

In dry-cleaning shops E = 1-2 of trichloroethylene exposure was observed during the fifties (Andersson, 1957). In a recent investigation of dry-cleaning shops E = 1-4 of tetrachloroethylene exposure was reported (Andersson et al., 1981). Today's limit values for 8-hour exposure (Anon., 1984 a) have been used in these and the following calculations.

In mechanical engineering shops E = 1-2 of trichloroethylene exposure was observed during the fifties (Andersson, 1957). The exposure on the average has decreased considerably during the seventies (Lindh et al., 1984), and is probably in most cases in compliance with today's limit values.

In the paint industry the exposure to organic solvents has decreased in general during the seventies, but manual cleaning of bulky pieces of equipment and big vessels with organic solvents is still a difficult operation. Very high exposures to a complicated mixture of organic solvents have been reported. E = 10-100 or even more was often found (Ulfvarson, 1977).

It has been suggested that the cleaning operations should be performed using organic solvents with low vapour pressures. The exposure can be roughly estimated by a combination of equations 1, 3 and 5 assuming no spot ventilation and a general ventilation of 30 m^3 per h per employee and an air velocity over the surface of organic solvent of 0.5 m/s. The exposure E_d at a certain diameter d cm is

$$E_d = 2 \cdot 10^{-5} \cdot d^{1.89} \cdot E_s \qquad \ldots\ 10$$

and for d = 100 cm

$$E_{100} = 0.12 \cdot E_s \qquad \ldots\ 10a$$

Only with a few glycols and glycol ethers can manual cleaning of bulky pieces of equipment be performed safely without personal protection such as a respirator with a chemical cartridge. With a thinner containing for instance a mixture of alcohols, acetates, ketones and aromatic hydrocarbons the average exposure in the working space will be around E = 50 according to equation 10 and of course it will be still higher close to the surface of organic solvents. This calculation is well borne out by observations.

In printing offices the exposure to toluene was formerly very high. Qualitative observations during the fifties indicated that workers were continuously dizzy from the exposure (Övrum, personal communication 1982). In the middle of the seventies E = 2-3 of toluene was observed in a photogravure printing office (Övrum et al., 1977). According to equation 10 this corresponds to the exposure from an open vessel containing toluene and having a diameter of 20-25 cm, provided the general ventilation of 30 m^3 per h per employee is assumed.

The application of paints with a spraygun or by dipping in a dip tank may be done in a safe way if the local exhaust ventilation is in order. Painting with a brush or roller on the other hand is a difficult operation from the work environment point of view. It may cause a high occupational exposure, especially if the painter or members of other professions must stay in the workroom for a long while during the process of drying. In the beginning, the exposure will emanate only from the open can with paint and the paint in the brush or roller, but after a while the wet painted surface will cause an increasing concentration of the organic solvents used in the product.

In an experiment, the concentration of white spirit was measured during and after the painting of 25 m^2 of board with a roller in a room with a floor area of 20 m^2 and a height of 2.5 m (Hallin, 1975). The ventilation was 100 m^3/h, corresponding to 5 m^3 per h per m^2. The painting operation took 12-21 minutes and within this time $E = 3$ was sometimes reached. If the paint was applied with a spraygun twice as high an exposure was reached in the same time. 10-15 hours later $E > 1$ was still observed.

SAMPLING STRATEGY IN EXPOSURE MEASUREMENTS

Let us now consider the properties of the organic solvents from the standpoint of sampling strategy. The uptake in man of various organic solvents from polluted air has been studied experimentally and reported in quite a few publications (Åstrand and Övrum 1976, Åstrand et al. 1976). The experiments have confirmed that organic substances used as solvents behave more or less in accord with simple physical laws. If the substances have a low solubility in blood, they will soon come close to a kind of equilibrium in which the blood concentration levels out and further uptake of the substance almost ceases. On the other hand if the solubility in blood is high, such an equilibrium will not be attained during the experiment. The uptake goes on and on or will tend to cease only after an exposure period of considerable length.

Generally speaking, non-polar substances have low solubility in water or blood and they will come close to equilibrium with the body during a working day involving exposure to a steady concentration in the air. Polar substances, especially if they have a low volatility, will have a high solubility in water or blood and the theoretical time to equilibrium is long. In practice equilibrium will never be attained, but there will probably be a certain accumulation effect, from one working day to the next, especially if the storage in fat tissues is considered.

The exposure to organic solvents during a working day will usually vary considerably. Intervals with low exposure will alternate with peak exposures. Substances which rapidly attain equilibrium will readily show increase or decrease in the body concentration resulting from this alternating exposure, while the slow substances will respond only to the average concentration during the day.

The situation is illustrated in Figure 4. In this figure the absorption of two substances in the body is illustrated when the average exposure is always 28 mg/m^3 (0.028 mg/l).

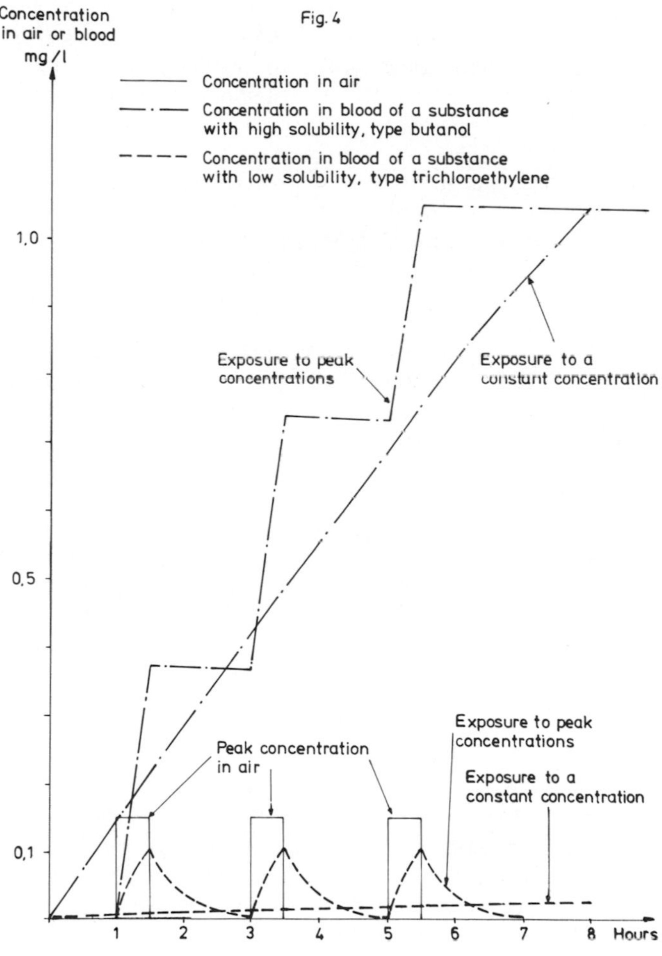

A simple model for the kinetics of uptake and elimination of anesthetic gases was used in the calculation (Henderson and Haggard, 1924). Solubility of the air pollutant in blood governs the uptake, which is modified by biotransformation of the substance. If biotransformation is rapid, the uptake will go on, since the blood will not be saturated. As already pointed out, organic solvents are characterized by a certain inertness and therefore as a first approximation in a general discussion of sampling strategy based on kinetics of uptake of organic solvents we may disregard biotransformation. Strictly speaking we should assume only acute effects, since the metabolites, if they are accumulating, may change the conclusions. One substance has a low solubility in water (or blood), partition coefficient assumed to be = 1. Trichloroethylene has a solubility in this range, and may be mentioned as an example although its effects are both acute and prolonged due to its metabolites. The other substance has a high solubility in water (or blood), partition coefficient assumed to be = 1000, and may be exemplified with butanol. In one application the average concentration is attained by three half-hour peaks of 150 mg/m^3 (0.150 mg/l) according to the figure, the rest of the time the concentration is 0. In a second application there is a constant concentration the whole working-day of 28 mg/m^3 (0.028 mg/l). The outcome after 8 hours' exposure will be quite different for the two substances. The substance with a low solubility will have a blood concentration of almost 0 in the first application and 0.028 mg/l in the second. The substance with a high solubility will have a blood concentration after 8 hours' exposure of 1.1 mg/l in both cases.

The conclusion is that we should adapt the sampling strategy to the properties and the effects of the substance we sample. For substances with mainly acute effects, such as impairments of the central nervous system, a sampling strategy which will register peak exposures is adequate if the water (blood) solubility is low, while average exposures should be registered for substances with a high water (blood) solubility.

REFERENCES

Anonymous (1973-74).Distribution of Wealth and Income, The new Encyclopaedia Britannica, 15th Benton, Published 1973-1974.

Anonymous (1984). Manufacturing 1982, Part 1 and Part 2. Official Statistics of Sweden, Publ by Statistics Sweden, Stockholm.

Anonymous (1984 a). Limit values for air contaminants. Swedish National Board of Occupational Safety and Health 1984:5. Stockholm (in Swedish).

Andersson A (1957). Gesundheitliche Gefahren in der Industrie bei Exposition für Trichloräthylen. Acta Med Scand (Suppl) 323:220.

Andersson I, Bornberg S, Seldén A (1981). Kemtvättprojektet 80/81. Rapport från yrkesmedicin 63/81, Regionsjukhuset i Örebro (in Swedish).

Carson B D (1976). Controlling vapour from open surface vessels. A Occ Hyg 19 313-324.

Hallin N (1975). Arbetshygieniska problem vid måleriarbete - lösningsmedelsångor från olika färgprodukter. Rapport från Bygghälsan Byggförlaget, Stockholm.

Henderson Y, Haggard HW (1924). Noxious gases. Renhold Publ Corp., New York.

Johansson L, Kulander K-E and Szabó (1984). Handling of solvents in the furniture industry and in wood impregnating. IVL-publ B 763, Stockholm (in Swedish).

Lindh T, Ulfvarson U, Vesterberg O (1984). Trends over several decades in excretion of trichloroacetic acid after occupational exposure to trichloroethylene. International Conference on Organic Solvent Toxicity, Stockholm. Arbete och Hälsa 1984:29, Arbetarskyddsverket.

Nylén P, Sunderland E (1965). Modern surface coatings. Interscience Publ, John Wiley & Sons Ltd.

Olander L (1982). Ventilation. Studentlitteratur, Lund (in Swedish).

Ulfvarson U (1977). Chemical hazards in the paint industry. International symposium on the control of air pollution in the working environment. Stockholm 6-8 Sept 1977, Part II. Solvents - Welding, pp 62-75. International Labour Office, Geneva, and Worker's Protection Fund, Stockholm.

Åstrand I, Övrum P (1976). Exposure to trichloroethylene. I. Uptake and distribution in man. Scand j work environ & health 4(1976) 199-211.

Åstrand I, Övrum P, Lindqvist T, Hultengren M (1976). Exposure to butyl alcohole - uptake and distribution in man. Scand j work environ & health 165-175.

Övrum P, Hultengren M, Lindqvist T (1977). Exposition för toluen och upptag i kroppen vid arbete i ett djuptryckeri. Arbete och Hälsa 1977:4, Arbetarskyddsverket, Stockholm (in Swedish).

Safety and Health Aspects of Organic Solvents, pages 21–30

SOLVENT CONTAINING PROCESSES AND WORK PRACTICES: ENVIRONMENTAL OBSERVATIONS

Pentti Kalliokoski

Department of Industrial hygiene, University of Kuopio, P.O.B. 6, 70211 Kuopio 21, Finland

INTRODUCTION

Organic solvents are volatile liquids used to dissolve resins and other materials for the ink, paint and coatings industries, as reaction solvents for the process industry, or as cleaning agents for the maintenance or dry-cleaning industries. These compounds include aromatic and aliphatic hydrocarbons, alcohols, ketones, esters, ethers, glycols, glycol ethers and organic halides. The physical properties of common organic solvents and their annual production rates in the United States are summarized by Parrish, (1983).

The situations where workers are exposed to organic solvents are practical to divide into manual and process operations (Kalliokoski, 1983). Manual emissions are caused by the worker due to the tasks he is performing. Those also include the emissions from small machines if the operator is exposed to them directly. Process emissions do not take place in the permanent or foremost work sites but from several sources of large processes. They include leakages from pumps, seams, and enclosure openings.

Most painting, gluing, metal degreasing and dry cleaning tasks can be classified as manual operations whereas the use of solvents in chemical, pharmaceutical and printing industries generally causes process emissions.

Petroleum refinery streams consists largely of compounds which could be classified as organic solvents but which are mainly used as fuels or as feedstock for petrochemical industry. The petrochemical production which does not involve the isolation of separate solvents may exceed the production of isolated solvents considerably. For example, only 17 % of toluene produced in the US is isolated, but is used in gasoline (Hoff, 1983). In addition, the actual solvent use may be only a minor fraction of the total production of the solvents. Toluene is again a good example; its annual use as a solvent is 379 000 t or only ca. 11% of the toluene produced in the US (Hoff, 1983).

Many of the classic solvent poisonings were caused by such especially toxic solvents as benzene and carbon tetrachloride. Those were replaced by safer solvents during the 1960s at least in the Nordic countries. Today, exposure to the most toxic solvents is rare even in laboratories. In the 1970s, also the use of other solvents began to decline because of occupational health and environmental considerations. This descent is expected to continue, and the solvent-based products will be substituted by fully solid or water-based formulations.

MANUAL EMISSIONS

Painting

Painting operations comprise the largest group of tasks where workers are exposed to organic solvents in most countries. Traditional oil paints for construction industry were replaced by solvent-based formulations in the 1960s. At that time, construction painters' exposure to solvents was at its height. Airborne solvent concentrations exceeded the current Finnish occupational health standards 4-5 times during painting and lacquering using epoxyester formulations (Riala and Kalliokoski, 1980).

Today water-based latex paints are most common in construction painting. The remaining solvent-based paints are mainly of the alkyd type and contain solvent naphtha as the primary solvent. It is much less volatile than the solvents used in most earlier paints. A study on the exposure to solvents in construction and maintenance painting showed that the present average exposure levels are low (about 40 ppm) in Finland. However, the construction painters still occasionally perform tasks in which the solvent concentration is excessive (up to 300 ppm of solvent naphtha; Riala et al., 1984).

In industrial painting, the solvent-based products are still the most usual type. The solventless powder paints are, however, becoming more common in metal industry. The exposure to solvents caused by industrial painting operations can often be controlled effectively by ventilation. NIOSH has carried out a comprehensive evaluation of engineering control technology for spray painting (O'Brien and Hurley, 1981). This survey indicates that exposure to organic solvents is readily controlled to less than 25 % of OSHA limits (about 40 % of the Finnish Occupational health standards) when spray booth ventilation meets the OSHA requirements. Occupational health limits for paint mist were exceeded more often than those for solvents. On the other hand, high concentrations of organic solvents were found in paint mixing and storage rooms. Similar observations have been made in Finland. The combined solvent exposure levels were usually below the old Finnish occupational health standard (given in 1972; correspond approximately to the OSHA limits) during actual painting situations already in 1968-1972 whereas all concentration levels detected in paint mixing and storage rooms exceeded the standards (Kalliokoski, 1975).

When transportation equipments, such as automobiles and railroad cars, are finished the painter must enter the spray booth otherwise than in the previous examples; therefore respirators should be worn during the work. Measurements done in the US during airless spray painting of railroad cars in the semi-downdraft spray booth revealed very high paint mist (27-72 mg/m^3) and solvent (combined exposure 96-199 % of OSHA limits) concentrations (O'Brien and Hurley, 1981).

Chemical exposures during automobile painting were recently studied in Finland. The mean combined solvent exposure was about half of the limit based on the current Finnish standards (Raatikainen et al., 1983). Even though the solvent levels were quite low, the respirators should also be worn during auto finishing because very high paint mist concentrations (mean 43.3 mg/m^3) were found. Quite similar results were obtained in earlier measurements. The solvent levels observed during painting of small automobiles between 1968 and 1972 were usually less than the occupational health standards, but during finishing of large vehicles the concentrations were high (Kalliokoski, 1975).
In another earlier Finnish study, the mean combined

airborne level of the solvent mixture was 31.8 % of the Finnish occupational health standards (given in 1972; Hänninen et al., 1976). It should also be noted that polyurethane paints are nowadays common in vehicle finishing. In the abovementioned most recent Finnish study, 13 % of the hexamethylene diisocyanate (HDI) measurements conducted in the painters' breathing zone exceeded the current Finnish standard. In addition, very high levels of HDI oligomers were detected (Raatikainen et al., 1983).

High airborne concentrations of solvents have been observed during spray painting in Finnish docks. In the measurements conducted between 1968 and 1972 the combined solvent exposure levels ranged from 90 to 365 % (of the standards given in 1972; Kalliokoski, 1975).

The exposure to solvents during spray finishing operations in furniture and other wood product industry can be effectively controlled by spray booths. The combined exposure levels did not exceed 25 % of OSHA limits in wood furniture finishing in the US (O'Brien and Hurley, 1981). Similar low concentrations were also detected in the Finnish wood furniture plants already in early 1970s (Skyttä, 1978). High levels were, however, occasionally found when painting is performed in poorly ventilated areas outside the booths and in paint mixing and storage rooms (Kalliokoski, 1975). The results of more recent measurements done by the Institute of Occupational Health confirm that heavy exposure to solvent vapors is rare in this industrial field (Kokko, 1982).

Formaldehyde containing resins are common in paints and lacquers used in furniture industry. A recent Finnish study revealed that exposure to formaldehyde is still a serious problem in this industry. As many as 42 % of the measurements exceeded the Finnish Occupational health standard of formaldehyde (1 ppm). The corresponding percentage was about 20 % for solvent measurements (Riipinen et al., 1985).

The amount of solvent measurements conducted by the Institute of Occupational Health during the various painting operations totalled 134 between years 1971 and 1976; 20.2 % of them exceeded the (year 1972) Finnish Occupational health standard. The noncompliance percentage was as high as 52 % for airborne paint mist levels during the same period of time (Skyttä, 1978).

Gluing

The carpet layers are exposed to very high concentrations of solvent vapors if they use contact glues for floor coverings. As high combined solvent exposure levels as 1000 % (of the Finnish standards given in 1972) have been observed during such operations (Kalliokoski, 1975). Fortunately those glues have largely been replaced by ethanol- and water-based glues.

The use of solvent-based glues may cause heavy exposure to solvents also in other industries. High concentrations have been detected especially in the footwear industry. However, these exposures can usually be prevented by means of exhaust hoods (Kalliokoski, 1975).

Paint and printing ink manufacture

Extremely high exposures to organic solvents were common during the manual tasks of paint and printing ink manufacture still in early 1970's. The most serious exposures took place in the rooms where vessels and barrels were cleaned with solvents. When a worker wiped the inside of a container with a cloth moistened with highly volatile solvents his breathing zone concentration reached levels high enough to cause intoxication. Conversations with workers in workplaces revealed that this led occasionally to solvent addiction. Some workers began to expose themselves intentionally and even continued the misuse at home. High exposure levels were also observed during many other tasks. Almost all the solvent measurements conducted by the Institute of Occupational Health in the Finnish paint and printing ink plants in 1971-72 exceeded the present occupational health standards. About half of the samples were out of compliance (if compared to current standards) still in 1975-76 (Skyttä, 1978). Statistical data on the present exposure levels in the Finnish paint and printing ink industry are not available, but the situation is probably clearly better due to more advanced production and control technology.

Significant exposures to solvents are also reported to occur in the Swedish paint industry (Ulfvarson et al., 1976). Average solvent levels exceeded the Swedish occupational health standards during most operations studied. Highest sample concentrations were over ten times higher than the health standard.

Degreasing

Exposure to solvents during cold degreasing of metal parts is easily prevented by providing the tank with local exhaust ventilation and avoiding the use of readily volatile solvents.

In vapor phase degreasing, the solvent is heated to its boiling point. The degree of solvent exposure depends on the effectiveness of condenser and local exhaust system. The metal parts must be totally dry before removing from the degreaser. Burges (1981) has summarized extensive data (based on surveys of some 1000 degreasers) on the airborne levels of chlorinated solvents for open-top vapor degreaser operations in the USA. Typical concentrations were 100-400 ppm.

Dry cleaning

Perchloroethylene is the most commonly used dry-cleaning solvent. Even though perchloroethylene concentrations have not been reported separately for dry-cleaning operations in Finland, high airborne perchloroethylene levels seem to be rare, because only 6.7 % of all (254) perchloroethylene measurements conducted between 1977 and 1980 exceeded the health standard (Kokko, 1982).

Reinforced plastic production

Workers are often heavily exposed to styrene in the reinforced plastic plants, especially during manual lay-up operations. A world-wide summary of the industrial hygiene surveys conducted in reinforced plastic industry has been published recently by WHO (WHO, 1983). This report indicates that exposure levels of 100-200 ppm have been typical. However, it is possible to reduce the concentration of styrene below 50 ppm by means of carefully designed ventilation systems (Isakson, 1976; Kalliokoski, 1984; Todd and Shulman, 1984; Kalliokoski et al., 1985). In addition to styrene, acetone (used as a clean-up solvent) contributes to workers' solvent exposure in reinforced plastic industry. Acetone levels are usually somewhat lower than those of styrene

(Kallikoski, 1976; Todd and Shulman, 1984).

PROCESS EMISSIONS

Photogravure printing

Extensive studies on the exposure to toluene in photogravure printing plants have been carried out both in Finland and Sweden (Kalliokoski, 1979; Övrum et al., 1978). In the Swedish study, the mean exposure concentrations during a three weeks period ranged from 63 to 118 ppm. The corresponding range was 68-186 ppm in the Finnish study. The Finnish study also included a retrospective industrial hygiene survey, in which the previous exposure levels were estimated on the basis of the long-term toluene consumption statistics of the plants and the relationships observed between the exposure levels and toluene consumption rates. This indicated that the intensity of toluene exposure has probably not changed noticeably for about 20 years, because the gain obtained with ventilation improvement has been outweighed by increased production.

In the Finnish study, it was found that some workers stayed unnecessarily often or long times in areas where the concentration of toluene was very high (up to 3500 ppm). Covers of presses and toluene containers were also held open. This may be due to addiction development. It is possible that this kind of misuse is inconscious.

Petroleum and petrochemical industry

The processes of petroleum and petrochemical industry are closed and mainly located outdoors. Thus the hydrocarbon levels are low under normal conditions. Even the exposure during abnormal situations can be greatly reduced by isolating the areas having highest potential for exposure, such as pumps and compressors which may leak, from the workers. Principles for minimization of leaks, spills and other releases in the petrochemical industry has been presented e.g. by Brief and Lynch (1978) and Jones et al. (1984).

The most significant solvent exposures in modern petrochemical industry are probably due to benzene vapors during the loading of motor gasoline. Benzene concentrations on the level of 2 ppm have been reported (Sherwood, 1972).

SUMMARY

Even though there has been a shift toward water-based or fully solid systems, organic solvents still comprise a significant occupational health hazard. Fortunately, exposure levels can nowadays be effectively controlled by proper enclosures and ventilation in most remaining applications of organic solvents, and, generally taken, the development of occupational health conditions has been favorable on the workplaces using organic solvents. When as many as 24.2 % of the 2639 solvent measurements carried out by the Institute of Occupational Health in Finland exceeded the occupational health standards between 1971 and 1976, such non-compliance levels were detected only in 3.0 % of the 2823 samples taken between 1977 and 1980 (Skyttä, 1978; Kokko, 1982).

The persons dealing with occupational health problems in workplaces should also be aware of the possible existence of solvent misuse. This may not develop into the level of solvent sniffing, but into a milder addiction. The workers adopt working habits that cause unnecessary exposure. Repeatedly found exceptionally high concentration levels in biological exposure tests are an indication of a possible abuse.

REFERENCES

Brief RS, Lynch J (1978). Industrial hygiene engineering in the petrochemical industry. Am Ind Hyg Assoc J 39: 620-625.

Burgess WA (1981). Recognition of health hazards in industry. John Wiley & Sons, New York.

Hoff MC (1982). Toluene. In: Kirk-Othmer Encyclopedia of chemical technology. Third edition. Vol. 23. John Wiley & Sons, New York, pp. 246-273.

Hänninen H, Eskelinen L, Husman K, Nurminen M (1976). Behavioral effects of longterm exposure to a mixture of organic solvents. Scan j work environ & health 4:240-255.

Isaksson G (1976). Methods to reduce the exposure to styrene in the reinforced plastic industry. Stockholm: Sveriges Plastförbund, 79 pp (in Swedish).

Jones AL, Devine M, Janes PR, Oakes D, Western NJ (1984). Fugitive emissions of vapours from process equipment. British Occupational Hygiene Society, Technical Guide No. 3, Science Reviews Ltd. Northwood.

Kalliokoski P (1975). Occasionally excessive paint mist levels. Työ-Terveys-Turvallisuus n:o 6:20-23 (in Finnish).

Kalliokoski P (1976). The reinforced plastic industry - A problem work environment. Työterveyslaitoksen tutkimuksia 122, 130 pp. (in Finnish).

Kalliokoski P (1979). Toluene exposure in Finnish publication rotogravure plants. PhD Thesis. University of Minnesota, 260 pp.

Kalliokoski P (1983). Some examples of technical prevention of exposure to organic solvents. In: Berufliche Exposition gegenüber organischen Lösungsmitteln. Zentralinstitut für Arbeitsmedizin der DDR, Berlin, pp. 125-138.

Kalliokoski P, Koistinen T, Jääskeläinen M (1984). in: Järvisalo J, Pfäffli P, Vainio H. (Eds.) Industrial hazards of plastics and synthetic elastomers, Alan R. Liss, New York, pp. 279-286.

Kalliokoski P, Säämänen A, Ivalo L (1985). Eddy formation and the effectiveness of local ventilation for control of exposure to styrene. Ist International Symposium or Ventilation for Contaminant Control, Toronto.

Kokko A (1982). Industrial hygiene survey statistics, years 1977-1980. Katsauksia 59, Työterveyslaitos, Helsinki, 82 pp (in Finnish).

O'Brien DM, Hurley DE (1981). An evaluation of engineering control technology for spray painting. NIOSH Publication No. 81-121, Cincinnati, 118 pp.

Parrish CF (1983). Solvents, industrial. In: Kirk-Othmer Encyclopedia of chemical technology. Third edition. Vol. 21. John Wiley & Sons, New York. pp. 377-401.

Raatikainen O, Saarinen L, Kalliokoski P, Laukkanen P (1983). Exposure to isocyanates in automobile finishing. Final report, Project 76/81, The Finnish Work Environment Fund, 16 pp. (in Finnish).

Riala R, Kalliokoski P (1980). Study of concrete reinforcement workers and maintenance house painters. Part 5. Solvent exposure in maintenance house painting. Työterveyslaitoksen tutkimuksia 171, Helsinki (in Finnish).

Riala R, Kalliokoski P, Wikström G, Pyy L (1984). Solvent exposure in construction and maintenance painting. Scand j work environ health 10:263-266.

Riipinen H, Priha E, Korhonen K, Liius R (1985). Exposure to solvents and formaldehyde in wood product finishing operations. Tampereen aluetyöterveyslaitos, 13 pp. (in Finnish).

Scherwood RJ (1972). Evaluation of exposure to benzene vapour during the loading of petrol. Brit J industr Med 29:65-69.

Skyttä E (1978). Industrial hygiene survey statistics, years 1971-76. Katsauksia 17. Työterveyslaitos, Helsinki. 190 pp. (in Finnish).

Todd WF, Shulman SA (1984). Control of styrene vapor in a large fiberglass boat manufacturing operation. Am Ind Hyg Assoc J 45:817-825.

Ulfvarson U, Rosen G, Cardfelt M, Ekholm U (1976). Chemical hazards in paint industry. Uppdragsraport Dnr 4979/75. Arbetarskyddsstyrelsen, Stockholm, (in Swedish).

World Health Organization (1983). Styrene. Environmental Health Criteria 26. WHO. Geneva. 123 pp.

Övrum P, Hultengren M, Lindqvist T (1978). Exposure to toluene in a photogravure printing plant. Scand j work environ & health 4:237-245.

Safety and Health Aspects of Organic Solvents, pages 31–41

Survey on Organic Solvents in Various Products and Methods for Estimating Workplace Exposures

E. Lehmann (1), J. Gmehling (2), U. Weidlich (2)
(1) Bundesanstalt für Arbeitsschutz
Vogelpothsweg 50-52, D-4600 Dortmund 50
(2) Universität Dortmund, Abteilung Chemietechnik
Postfach 500500, D-4600 Dortmund 50

INTRODUCTION

Monitoring of workplace exposure to carcinogenic compounds has been mandatory in the Federal Republic of Germany since 1980. It is now intended to extend these regulations to cover all toxic substances for which the limiting exposure values are established.
Experience in the past five years has shown that the enforcement of monitoring programmes can fail because of incomplete knowledge of the composition of the chemicals used at the workplace. Also, labelling and safety data sheets are insufficient to provide the user with the information needed to estimate exposure hazards.

In the course of preparing a strategy for implementing these new regulations organic solvents were identified as a group of main concern because of their wide spread uses in many industries as constituents of various products.

In order to (i) set priorities in standardizing analytical methods for workplace monitoring of organic solvents and (ii) develop procedures for evaluating combined exposures, a survey on organic solvents present in a variety of commonly used products like thinners, degreasers, paints, inks, adhesives, reagents etc. was carried out. The aim of the study was to identify the most common organic solvents and combinations of chemical classes in solvent mixtures.

SURVEY ON ORGANIC SOLVENTS

The samples were collected by factory inspectors in 1978-1982 from workplaces in various industries and were analyzed by gas chromatography. In 275 products, the total number of solvents identified was 34. In addition, five solvent mixtures (high and low boiling petrol distillates, gasoline, high boiling oils, and turpentine) were found to be in common use.The solvents were defined to belong to seven chemical classes: aromatic hydrocarbons, esters, ketones, alkanes, alkohols, ethers and halogenated hydrocarbons (see Table 1).

Table 1 List of most common solvents in thinners, degreasers, paints, inks and some reagents

Chemical name of solvent	(Frequency)*
AROMATIC HYDROCARBONS	
Benzene	(0.4)
Toluene	(25.5)
Xylenes	(36.7)
Ethylbenzene	(26.5)
1,3,5-Trimethylbenzene	(1.1)
Styrene	(0.4)
ALCOHOLS	
Methanol	(9.8)
Ethanol	(2.9)
n-Butanol	(6.9)
iso-Butanol	(10.2)
iso-Propanol	(6.9)
Diacetone alcohol (4-Methyl-2-pentanone-4-ol)	(0.7)
ALKANES	
n-Hexane	(0.7)
n-Heptane	(0.4)

Table 1 continued

ESTERS	
Methyl acetate	(3.7)
Ethyl acetate	(12.0)
Butyl acetate	(16.0)
Isobutyl acetate	(1.8)
Ethylene glycol monomethyl ether, acetate	(0.7)
Ethylene glycol monoethyl ether, acetate	(3.4)
Ethylene glycol monobutyl ether, acetate	(0.4)
ETHERS	
Ethylene glycol monoethyl ether	(1.9)
Ethylene glycol monobutyl ether	(0.4)
CHLORINATED HYDROCARBONS	
Dichloromethane	(6.9)
Trichlorofluoromethane	(0.7)
1,1,1-Trichloroethane	(6.9)
1,1,2-Trichlorotrifluoroethane	(1.5)
Trichloroethylene	(4.7)
Perchloroethylene	(5.1)
Pentachlorophenol	(0.4)
KETONES	
Acetone	(12)
Methyl ethyl ketone	(6.2)
Methyl isobutyl ketone	(3.3)
Cyclohexanone	(1.1)
SOLVENT MIXTURES	
Low boiling petrol distillate fractions	(27.6)
High boiling petrol distillate fractions	(23.3)
Gasoline	(3.6)
High boiling oils	(1.8)
Turpentine	(0.4)

* (Numbers observed/ Numbers analyzed) x 100

The most frequent combinations of solvent classes found in the formulations are listed in Table 2.

Table 2 Most common solvent class combinations

Combinations	Numbers
Aromatic hydrocarbons - Esters	15
Aromatic hydrocarbons - Chlorinated Hydrocarbons	5
Aromatic hydrocarbons - High and low boiling petrol distillatefractions	15
Aromatic hydrocarbons - Esters - Alcohols	10
Aromatic hydrocarbons - Esters - Ketones	11
Aromatic hydrocarbons - Alcohols - Ketones	3
Aromatic hydrocarbons - Esters - High and low boiling petrol distillate fractions	2
Aromatic hydrocarbons - Ketones - High and low boiling petrol distillate fractions	4
Aromatic hydrocarbons - Halogenated Hydrocarbons - High and low boiling petrol distillate fractions	7
Aromatic hydrocarbons - Esters - Alcohols - Ketones	8
Aromatic hydrocarbons - Esters - Alcohols - Ketones - High and low boiling petrol distillate fractions	3

Although several hundred organic solvents with industrial uses are mentioned in technical reports only about 15 of them seem to be present at most workplaces with the solvent exposures. The aromatic hydrocarbons xylene, toluene and ethylbenzene are the most common solvents, followed by the esters ethyl acetate and butyl acetate. Toluene was found in combination with esters and/or ketones, xylene (mainly the meta-isomer) with ethylbenzene and alcohols. Their contents in most products vary from a few percent to 50%. Halogenated hydrocarbons are very common, mainly in degreasers, where they are usually detected as a single component. Degreasers containing 1,1,1-trichloroethane may also contain 1,4-dioxane as an additive. Some solvents are typical for one product class, for example n-hexane used in adhesives, or petrol distillate fractions used in thinners and degreasers.

The frequencies obtained were based on a small number of samples and may not be representative. A Japanese survey based on 1179 samples gave similar results concerning the main solvents used in these classes of products. The situation seems to be comparable in industrialized countries (Inoue, 1983).

ESTIMATION OF WORKPLACE EXPOSURES

The typical compositions of thinners, degreasers, paints, inks and adhesives may give a good estimate of exposure risk at the workplace, if they can be correlated with workplace concentrations. The survey has shown that exposure to solvents such as toluene, xylenes, ethyl acetate etc. are evidently most common. As Figure 1 demonstrates, the probability of being exposed

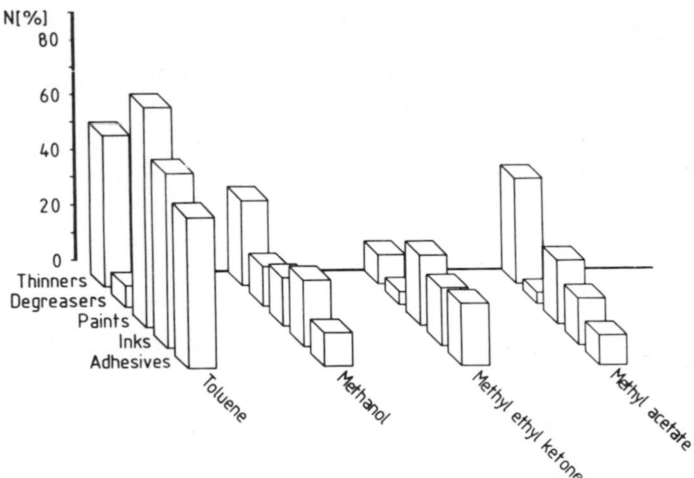

Fig. 1: Frequencies of four most common solvents in product classes

to a given solvent may differ very strongly within the product classes. The extent of exposure, however, depends not only on the composition of the product but also on the specific conditions of use.

In order to predict airborne concentrations quantitative data on the evaporation and dispersion of the solvent into the workplace environment are needed. For this purpose a concept was developed which enables us to calculate workplace concentrations for constituents of mixtures, if some physical/chemical data of the pure substances are available. The algorithm used is based on two theoretical models (Gmehling, 1984).

1. The two-film theory describing mass transfer across the phase boundary
2. The model of an ideal continously stirred vessel describing the residence time distribution in a ventilated room.

The amount of substance $\dot{n}_i$ transferred across an area F by evaporation is given by the following equation:

$$\dot{n}_i/F = k_{g,i}(p_{i,eq} - p_i)/RT$$

$k_{g,i}$ = overall mass transfer coefficient (related to the gas phase) [m/h]
$p_{i,eq}$ = equilibrium partial pressure of the component i [Torr]
p_i = partial pressure of the component i [Torr]
R = gas constant [Torr m^3/K mol]
T = temperature [K]

The total resistance $1/k_{g,i}$ is the sum of two partial resistances:

$$1/k_{g,i} = 1/ß_{g,1} + H^c_{i,j} /R\ T\ ß_{l,i}$$

$ß_{g,i}$ = gas phase mass transfer coefficient [m/h]
$ß_{l,i}$ = liquid phase mass transfer coefficient [m/h]

$H^{c}_{i,j}$ = Henry's constant referring to concentration [Torr m^3/mol]

PHASE EQUILIBRIUM

The phase equilibrium and the Henry's constants are calculated using the group contribution method UNIFAC (Fredenslund, 1977). This method permits the calculation of activity coefficients as a function of temperature and composition.

The activity coefficients γ_i of the components of a liquid mixture describe the deviation from ideal behaviour, i.e. the various interactions in the liquid phase. Apart from the molar concentration, the van der Waals volumes and surface areas, which can be obtained from the tables only the group interaction parameters (Fredenslund, 1977; Gmehling, 1982) must be known. The latter describe the interaction between the various subgroups of all components involved and can be obtained from experimental vapor liquid equilibrium data.

For the calculation of Henry's constant the following equation applies:

$$H_{i,j} = \gamma_i^{\infty} P_i^{s}$$

γ_i^{∞} = activity coefficient of the component i at the infinite dilution in solvent j

P_i^{s} = saturation vapor pressure of component i (available from the Antoine equation) [Torr]

The calculation of the amount of compound evaporated requires information on the driving force, the pressure gradient $p_{i,eq} - p_i$. The equilibrium partial pressure can be obtained from the following simplified equation:

$$p_{i,eq} = x_i \gamma_i P_i^{s}$$

x_i = mole fraction of component i

γ_i = activity coefficient of component i

MASS TRANSFER COEFFICIENT

A comprehensive search of the literature revealed that only very few data have been published and that these differ very greatly in some cases. Mackay (1973) has expressed the result of his experimental studies in the form of the following equation for calculating the gas phase mass transfer coefficient:

$$\beta_g = 0.0292\ v^{0.78}\ x^{-0.11}\ Sc^{-0.67} \quad [m/h]$$

v = air velocity [m/h]
x = diameter of the vessel from which evaporation occurs [m]
Sc = Schmidt number

In order to obtain more exact information mass transfer coefficients were experimentally determined for some of the most common solvents using the pure substances. A linear dependence of the amount of substance evaporated was found at air velocities between 0.2 to 0.7 m/s, which are typical for workplace situations. For rough estimations the mass transfer coefficient can be treated as a constant at a given air velocity within this range. At higher air velocities, however, $\beta_{g,i}$ considerably increases (Fig. 2).

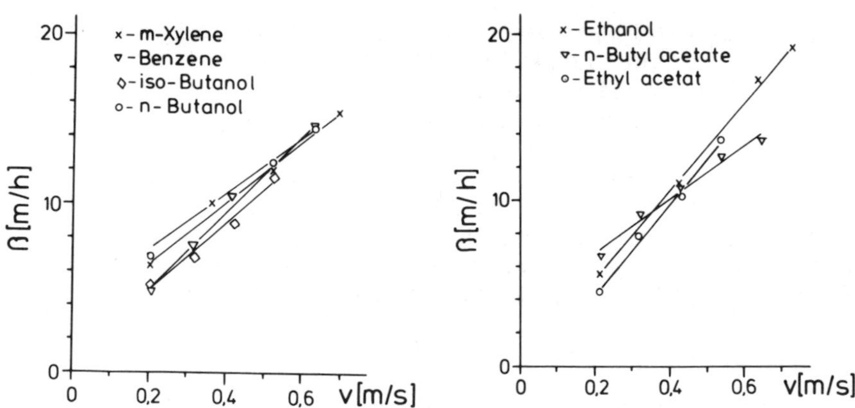

Fig. 2: Dependence of ß on air velocity

Measurements of solvent mixtures revealed that the contribution of the liquid phase mass transfer coefficient can be neglected.

The experimental data were then used to adapt the following equation (Colburn,1933) for predicting mass transfer coefficients at air velocities in the range of $0.2\ m/s < v < 0.7\ m/s$.

$$Sh = a\ Re^b\ Sc^c$$

Sh = Sherwood number
Re = Reynolds number

CALCULATION OF WORKPLACE CONCENTRATIONS

Once the amount of solvent set free is known, its concentration in the workplace environment can be calculated using parameters specific to the workplace. For this purpose the residence time distribution of an ideal continously stirred vessel is applied (Müller,1961). This simplified model initially assumes that after evaporation the solvent becomes completely mixed with ambient air. If a ventilation system is present, the stream of fresh or circulated air can be taken into consideration, if necessary with a residual concentration.

Algorithms were derived to calculate workplace concentrations for the following emission situations (Gmehling,1984):

- continous emission
- intermittent emission
- emission at a single rate
- emission at a decreasing rate.

The description at the continuous emission follows from this fundamental equation.

$$c = ((\dot{n} + (1-U)\,\dot{V}c_R) / ((1-U)\dot{V}))\,(1-\exp(-(1-U)\dot{V}t/V))$$

c = concentration [g/m^3]
$\dot{n}$ = mass transfer of component i [g/h]
U = ratio re-circulated air/fresh air
c_R = residual concentration in fresh air [g/m^3]
$\dot{V}$ = fresh air [m^3/h]
V = workroom size [m^3]
t = time [h]

It must be noted that the concentration is calculated as a function of time. It is known from flow patterns which have been obtained from experiments using smoke that extreme local concentrations can occur. The concentration given by our model can thus be taken as average concentration and can serve as a basis for judging primarily mean long-term exposures at the workplace.

REFERENCES

Colburn, A.P. A method of Correlating Forced Convection Heat Transfer Data and Comparison with Fluid Friction, AIChE J. (1933) 29, p 174

Fredenslund, A., Gmehling, J., Rasmussen, P. Vapor-Liquid Equilibria Using UNIFAC Elsevier, Amsterdam (1977)

Gmehling, J., Rasmussen, P., Fredenslund, A. Vapor-Liquid Equilibria by UNIFAC Group Contribution.
Revision and Extension. 2
Ind.Eng.Chem.Process Des.Dev. (1982) 21, p 118

Gmehling, J., Schwaitzer, U.
Berechnung von Expositionen beim Umgang mit lösemittelhaltigen Zubereitungen, FB 382, Ed. Bundesanstalt für Arbeitsschutz Dortmund, Wirtschaftsverlag NW, Bremerhaven (1984)

Inoue, T. et al. A Nationwide Survey on Organic Solvent Components in Various Solvent Products
Part 1: Industrial Health (1983) 21, pp 175-183
Part 2: Industrial Health (1983) 21, pp 185-197

Mackay, D., Matsugu, R.S.
Evaporation Rates of Liquid Hydrocarbon Spills on Land and Water
Can.J.Chem.Eng. (1973) 51, p 434

Müller, K.-G.
Bestimmung der erforderlichen Zuluftmenge bei lufttechnischen Anlagen
Heizung-Lüftung-Haustechnik (1961) 12, p 216

TOXICOKINETICS AND MECHANISMS OF TOXICITY

Safety and Health Aspects of Organic Solvents, pages 45–60

UPTAKE AND DISTRIBUTION OF COMMON INDUSTRIAL SOLVENTS

Karl-Heinz Cohr, M.Sc.

Department of Chemistry and Toxicology, National Institute of Occupational Health, Baunegårdsvej 73, DK-2900 Hellerup, Denmark.

INTRODUCTION

The toxic action of a chemical is related to the concentration of the chemical on the site of its action. The active toxic component may be the parent chemical and/or a biotransformation product. The concentration of the chemical substance at the site of action is dependent on its toxicokinetics i.e. absorption, distribution to and uptake into tissues, biotransformation and excretion.

Organic solvents may be taken up into the organism via all the common routes of exposure. During occupational exposure, uptake by inhalation makes out the most important route. Uptake through the skin may be of importance. However, this route is often overlooked. After uptake the solvents are distributed with the blood to the tissues and organs of the body, where they are accumulated and/or biotransformed. Organic solvents are mainly excreted via the lungs or in the urine (figure 1).

BASIC TOXICOKINETIC FEATURES

The uptake and distribution of a chemical is governed mainly by its biosolubility and the rates at which it is carried to the body and its tissues. The biosolubility of a compound is primarily composed of its lipid solubility (lipophilicity) and its water solubility (polarity, hydrophilicity). In general, lipophilic organic solvents are hydrophobic, i.e. they are only little soluble in water.

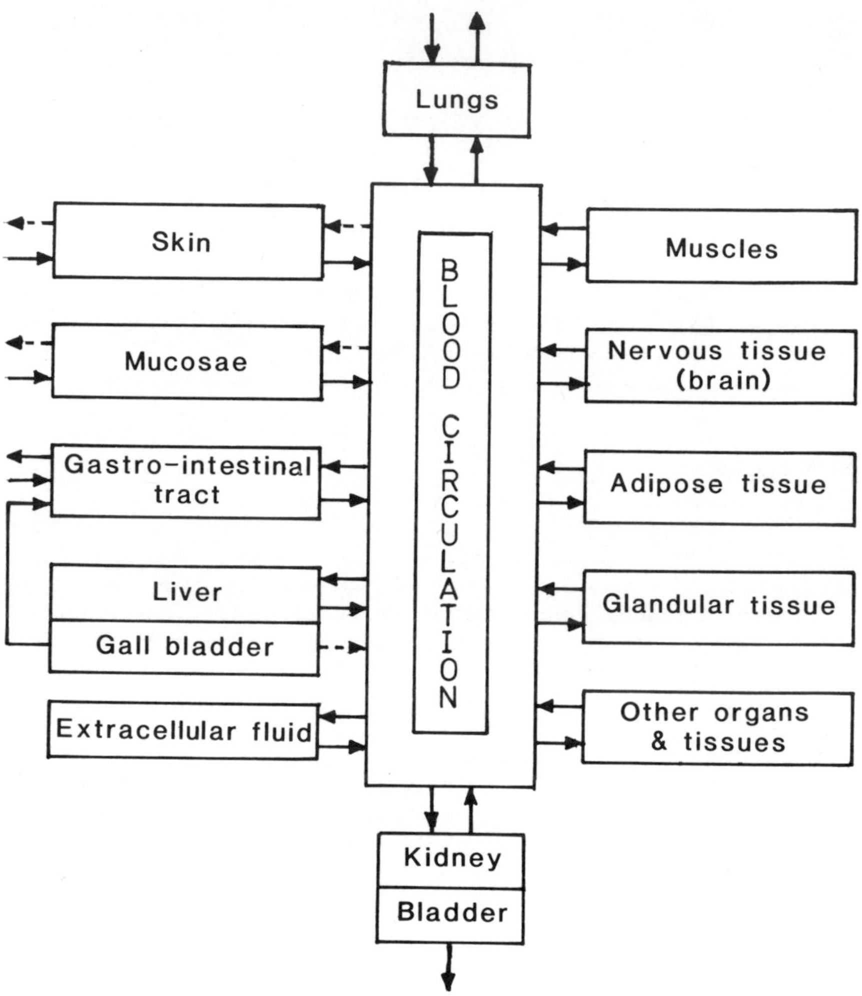

Figure 1. Schematic toxicokinetic model

However low molecular weight substances containing polar groups with Nitrogen or Oxygen atoms are partly soluble in or even miscible with water. Solvents which are both lipophilic and hydrophilic have special biological properties.

The biosolubility is commonly expressed as the partition coefficient between tissue (or body fluid) and air. In table 1 values of water, blood and oil (fat) solubilities for some organic solvents are given as *in vitro* partition coefficients at 37°C.

TABLE 1. Biosolubilities of selected organic solvents

Compound	Partition coefficient (37°C)		
	Water/air	Blood/air	Oil/air
Toluene	2.2	15.6	1471
Xylene	1.7	26.4	3842
Styrene	4.7	52	5465
Dichloromethane	7.2	9.7	152
1,1,1-Trichloroethane	0.9	3.3	356
Trichloroethylene	1.7	9.0	900
Tetrachloroethylene	0.4	13.1	1917
Acetone	395	245	86
Butanone	254	202	263

The total amount of solvent, which may be accumulated in a tissue is determined by the product of the tissue volume and the solubility of the solvent in the tissue (tissue/blood partition coefficient). This product is often called the volume of distribution. Accummulation in a tissue continues until the partial pressure of the solvent in the tissue is equal to that in the blood (and in the alveolar air). The rate, at which this equilibrium is established, is proportional to the volume of distribution and inversely proportional to the blood flow rate to the tissue.

The tissues and organs of the body may be divided grossly into groups according to their blood supply (table 2). Furthermore table 2 shows tissue/blood distribution coefficients for groups of hydrophilic and lipophilic solvents. *In vivo* data come from animal exposure for a

TABLE 2. Ratios for distribution of solvents between tissues and blod

Tissue	Tissue volume (liter)	Blood flow (l/min)	Partition coeffients at (37°C) Lipophilic in vivo	Lipophilic in vitro	Hydrophilic in vivo	Hydrophilic in vitro
VRG	5.8	3.45	1.1–2.2	1.1–2.1	0.6–0.9	0.8–1.4
-brain	1.5	0.81	1.1–1.7	1.9	0.8	0.9
MG	36.1	1.69	0.5–0.6	1.2	0.6	1.2
FG	13.0	0.35	(45)	>100	0.2–0.3	0.9
VPG	8.2	0.05	-	-	-	-
Blood	5.4	6.35	-	-	-	-

Partition coeffients (in vitro):

	Lipophilic	Hydrophilic
fat/blood	>approx. 60	<approx. 5
blood/air	<approx. 100	>approx. 200

VRG: Vessel rich group, e.g. heart, spleen, liver, kidney, lungs, (brain).
MG: Muscles
FG: Adipose tissues.
VPG: Vessel poor group, e.g. bones, lung parenchyme.

shorter period of time. Saturation of tissues has not necessarily been attained, thus <u>in vivo</u> data values are lower than <u>in vitro</u> values. It appears that hydrophilic solvents are distributed evenly to all tissues, having tissue/blood distribution coefficients of approx. 1. Lipophilic solvents have distribution coefficients of 1–2 for most tissues, except, for the lipid-rich tissues, in which case there may be a large accumulation. The brain which is considered a target organ for the toxic action of solvents has distribution coefficients that are comparable to the remainder of the vessel rich group, althrough it is rich in lipids (grey and white matter). This may be due to the high blood content. However, the concentration of solvent in lipoid subcellular structures could be much higher, e.g. membranes.

Biotransformation of the solvent will lower the concentration in the tissues, and a state of equibibrium will be reached at a lower solvent concentration in the tissues. The biotransformation capacity may be influenced by other chemical compounds e.g. in the work environment, in the general environment, in the food or drugs, or by individual biological factors, e.g. age, sex or genetic constitution.

Elimination of unchanged lipophilic compounds takes place mainly via the lungs in a process reverse to the uptake in the lungs. Hydrophilic solvents and biotransformation products are primarily excreted in the urine.

UPTAKE VIA THE LUNGS

Vapours of organic solvents are inhaled into the alveoli during respiration. In the alveoli the vapours diffuse into the capillary blood. This diffusion is fast due to small distances, and an equilibrium between the concentrations in the blood and in the alveolar air is rapidly established. The amount taken up depends on the vapour concentration in the alveolar air, the solubility in blood, the solvent concentration in mixed venous blood returning to the lungs as well as the pulmonary ventilation and the cardiac output. The rate limiting factor for establishing

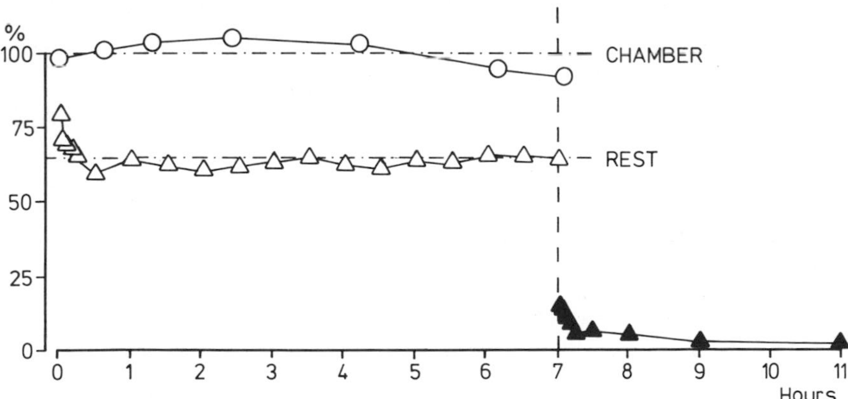

Figure 2. Relative uptake of aliphatic hydrocarbons during exposure to white spirit. Open circles are concentrations in inhaled air. Open triangles are relative uptake. Filled triangles are relative concentrations in end-exhaled air after exposure.

the equilibrium is the depth and rate of respiration in the case of a soluble solvent (blood/air partition coefficient larger than approx. 10), and the cardiac output in the case of less soluble solvents.

During constant environmental concentration and constant physical workload the uptake is fairly constant as seen from figure 2. Volunteers were exposed sedentary to white spirit for 7 hours. It is seen that a constant solvent concentration in end-exhaled air is rapidly attained and maintained throughout the exposure period. The area under the curve is a measure of the total amount taken up. At the end of exposure the concentration in alveolar air decreases rapidly. The figure indicates that only a minor amount of the retained solvent is exhaled after exposure. In fact it was calculated from the data that only 10% of the retained amount was exhaled during the next 19 hours.

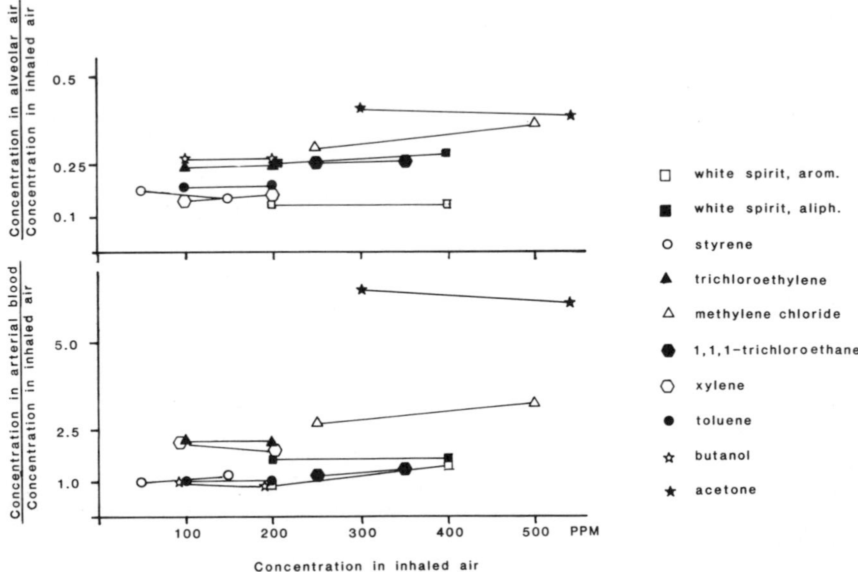

Figure 3. Relative concentration of some common organic solvents in alveolar air and in arterial blood after 30 minutes of exposure.

In figure 3 the results from 30-minute exposures of sedentary volunteers to different solvents are compiled. It may be seen that the relative concentrations in alveolar air and in arterial blood are constant for the single solvents regardless of the concentration in the environmental air. This is the case for lipophilic as well as for hydrophilic solvents. This also means that the relation of concentration in arterial blood to concentration in alveolar air is constant during such exposure conditions. The uptake is fairly constant and amounts to 50-85% of the solvent vapour present in alveolar air.

As the exposure extends the body accumulates solvent and the concentrations of solvent in blood and in alveolar air increase as equilibrium for the whole body is approached. The uptake of the lipophilic solvents relates to the solubility in blood; the least soluble (e.g. methylene chloride, 1,1,1-trichloroethane) being the least absorbed. The picture is not so unequivocal for the hydrophilic solvents (e.g. acetone, butanol). In this case there must be other or more factors governing the uptake.

Physical workload influences on the solvent concentration in alveolar air as may be from figure 4. As the workload increases the concentration in alveolar air increases, when speaking of the lipophilic solvents. Again there is a correlation to the solubility in blood. A physical workload increases the respiration rate as well as the pulse rate, the net result being an increased uptake. Apparently alveolar air concentration of hydrophilic solvents is not influenced by a physical workload. An explanation for this phenomenon may be that end-exhaled air is measured and not alveolar air. Hydrophilic solvents may be absorbed in the mucus layer covering the surface of the respiratory tract when inhaling and released again when exhaling.

The relation of the concentration in arterial blood to the concentration in alveolar air (i.e. end-exhaled air) is a function of physical workload. For solvents with blood/air partition coefficients larger than approx. 30 this relation increases with increasing workload (e.g. styrene, xylene, aromatics of white spirit, acetone). Those not influenced by a physical workload have a blood/air partition coefficients of less than approx. 20. (e.g. toluene, trichloroethylene, dichloromethane, 1,1,1-trichloroethane, aliphatics of white

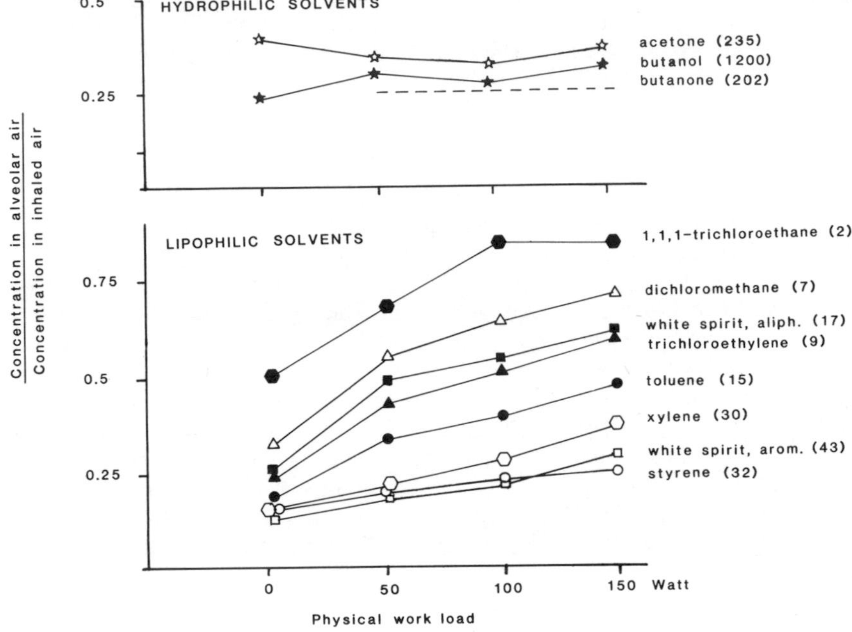

Figure 4. Relative concentration in alveolar air of some common organic solvents after 30 minutes exposures during a physical work load. Numbers in parentheses are partition coefficients.

spirit). Butanol having a partition coefficient of 200 does not fit into this pattern.

Åstrand et al have correlated the uptake of solvent as percentage of supplied amount to the alveolar air concentration as percentage of inhaled concentration (figure 5). For lipophilic solvents this relation depicts a straight line, which intercepts the x-axis at 100% and the y-axis at approximately 76%. The reason that the y-intercept is not 100% is due to the influence of the dead space in the respiratory tract. Acetone, which is a hydrophilic solvent follows this pattern too, whereas butanol, which also is hydrophilic, does not.

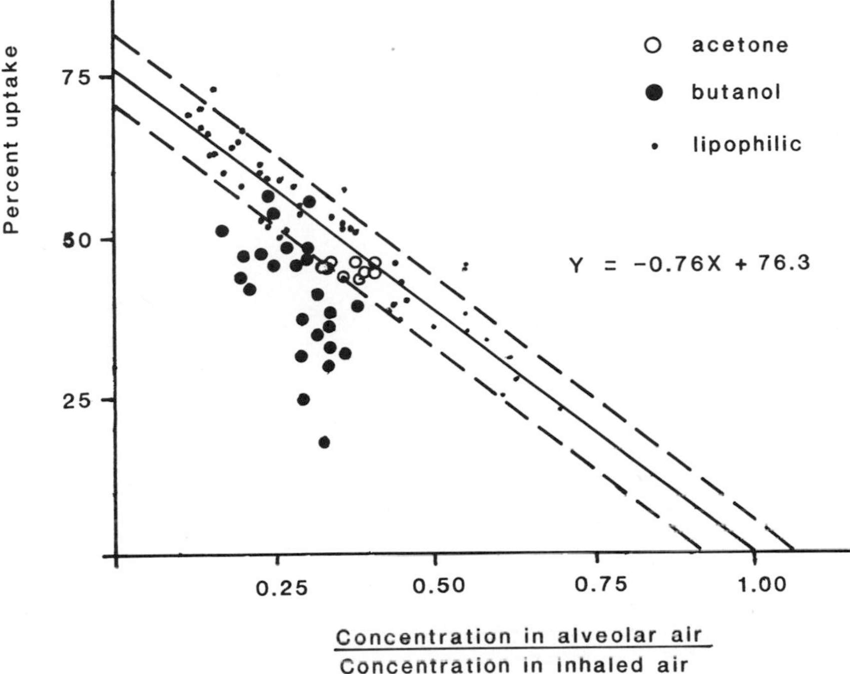

Figure 5. Uptake of organic solvents during exposure via the lungs, as percentage of the inhaled amount.

UPTAKE VIA THE SKIN

The skin acts as a barrier for many substances. It behaves in many respects as a lipoid-water sandwich. A consequence of this would be that substances which penetrate the skin should have a lipid-water partition coefficient of approx. 1. This is the case for e.g. lower ketones and alcohols. Penetration of the skin is typically biphasic. In the first phase some solvent presumably accumulates in the skin resulting in the build-up of a concentration profile. The build-up of the profile takes some time (i.e. lag time), which increases with increasing oil/water partition coefficient in a non-linear way. In the second phase the solvents penetrate the skin with a constant velocity. This velocity is related to the solubility in water. In a double logarithmic graph the relation is forming a straight line (figure 6).

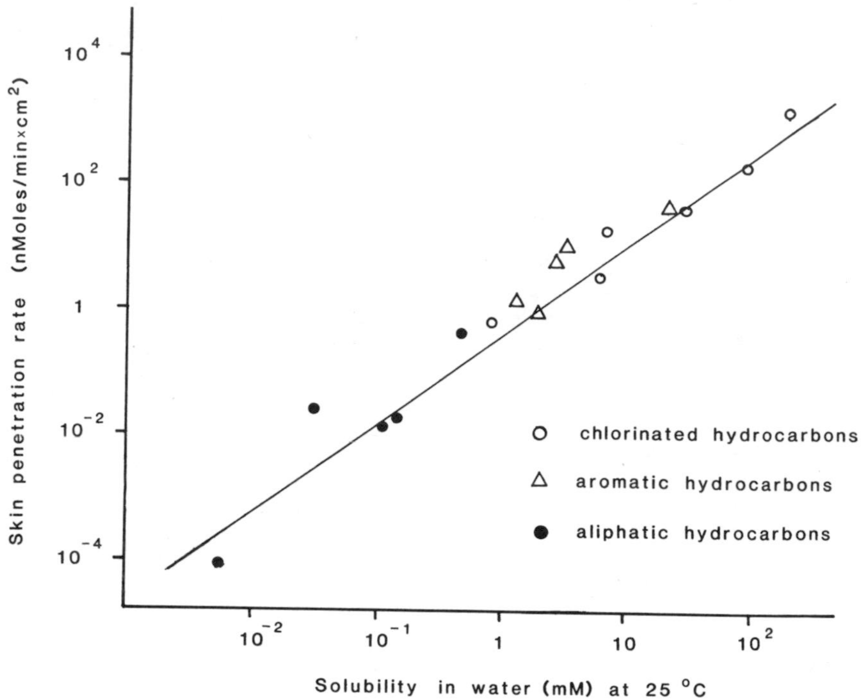

Figure 6. Skin penetration rates for some common organic solvents in relation to their water solubility.

When skin contact stops the solvent accumulated in the skin will disappear in one of two ways. Either by evaporation from the skin, as it is known from e.g. petrol, or by uptake in the body, as it is known for e.g. toluene and some chlorinated hydrocarbons. In the latter case, exposure will continue after skin contact has ceased.

It should be mentioned that uptake of solvent vapours through the skin is a possibility too. In an experimental study with human volunteers Riihimäki & Pfäffli found that one per cent or less of solvent vapour was taken up through the skin as compared to the amount retained in the body via the lungs for the same period of time.

DISTRIBUTION AND ACCUMULATION IN THE BODY

Exposure of human volunteers shows an accumulation of lipophilic solvents in adipose tissue. The concentration in adipose tissue depends on the obesity of the person. At equal exposure conditions obese persons will have lower solvent concentration in adipose tissue than lean persons. However, the total amount of solvent accumulated is positively correlated to the volume of the adipose tissue.

TABLE 3. Biological half lives of solvents in human adipose tissue

	Persons lean–obese (hours)	Partition coefficients fat/blood	blood/air
Toluene	12–65	94	15
Xylene	48–70	129	28
Styrene	53–96	105	51
White spirit aliph.	approx. 180	400	17
Acetone	–	0.4	245
Butanone	–	1.3	200
Biological half life:	$t_{1/2} = \ln 2 \cdot \lambda \cdot \frac{V}{F}$		

The biological half lives of solvents in adipose tissues have been shown to be longer for obese persons than for lean. This is not surprising, as the biological half life is proportional to the volume of the adipose tissue. Table 3 lists biological half lives determined from human subcutaneous fat biopsies for some solvents, and their corresponding fat/blood- and blood/air partition coefficients determined <u>in vitro</u>. It may be seen that there is a fairly good correlation between the <u>in vitro</u> determined fat/blood coefficients and the <u>in vivo</u> determined biological half lives.

The biological half life time ($t_{1/2}$) may be calculated from the formula in the lower part of table 3 if one knows the tissue/blood coefficients (λ), the tissue volume (V) and the blood perfusion rate of the tissue (F).

Exposure to solvents with biological half lives shorter than approx. 8-10 hours should theoretically not result in an accumulation in the body. Exposure to solvents with longer biological half lives at constant exposure conditions would result in attainment of an equilibrium after approx. 8 weeks of exposure.

This is illustrated in figure 7, which depicts exposure of volunteers to white spirit for 6 hours/day for 5 consecutive days. Already during the first week of exposure the concentration in the adipose tissue amounted to nearly half of the equilibrium concentration, which was calculated to approx. 85 mg/kg tissue.

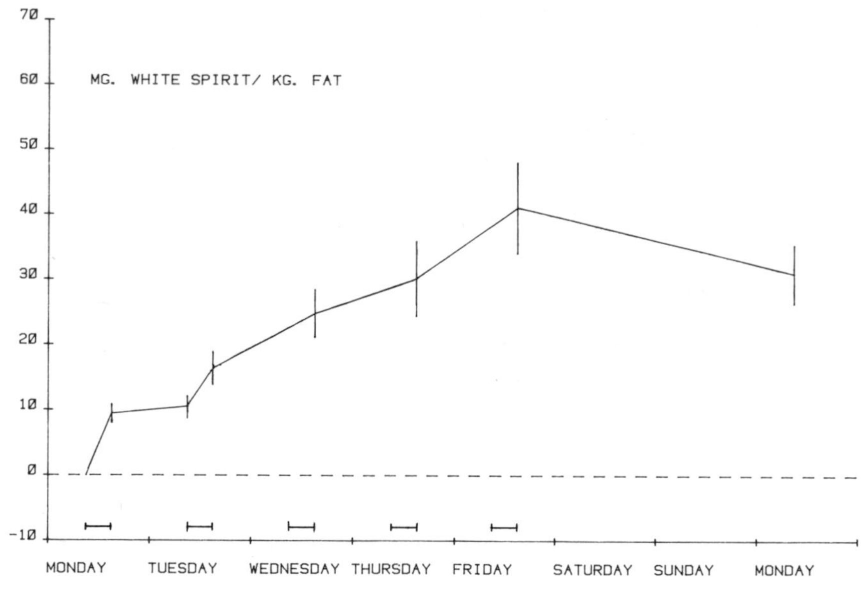

Figure 7. Accumulation of aliphatic white spirit in subcutaneous fat during exposure of human volunteers to 100 ppm for 5 consecutive days. Horizontal bars indicate the 6-hours exposure periods. Vertical bars indicate the standard deviation for 7 volunteers.

An accumulation like this would also occur in persons occupationally exposed to lipophilic solvents, meaning that workers will carry a body burden. Subcutanous fat biopsies from polymerisation plant workers show that these persons have styrene in their adipose tissue on monday morning before starting at work.

A body burden of solvents may pose a health risk due to redistribution from adipose tissue to e.g. nervous tissues.

Corresponding information on hydrophilic solvents is not available, but it does not seem likely that hydrophilic solvents should accumulate in the body, because of their fairly short biological half lives and their even distribution throughout the body.

CONCLUSION

To summarize it may be stated that organic solvents are taken up in the body by inhalation of vapours. The amount taken up being governed by the environmental concentration, the type of solvent, and the physical work load, i.e. pulmonary ventilation and cardiac output. At rest 50-85% of the inhaled amount is retained in the body. Some solvents may be absorbed through the skin.

Hydrophilic solvents distribute evenly throughout the body, whereas lipophilic may accumulate in the adipose tissues of the body.

REFERENCES

Basic toxicokinetic features

Cowles AL (1970). "The Uptake and Distribution of Inhalation Anesthetics". Dissertation from University of Rochester.

Fiserova-Bergerova V (ed) (1983). "Modelling of Inhalation Exposure to Vapors: Uptake, Distribution and Elimination". Boca Raton, Florida: CRC Press. Volumes I & II.

Kety SS (1951). The theory and application of the exchange of inert gas at the lungs and tissucs. Pharmacol Rev 3: 1-41.

Papper EM, Kitz RJ (eds) (1963). "Uptake and Distribution of Anesthetic Agents". New York: McGraw-Hill.

Uptake via the lungs

Carlsson A (1982). Exposure to toluene. Uptake, distribution and elimination in man. Scand J Work Environ Health 8: 43-55.

Carlsson A, Lindqvist T (1977). Exposure of animals and man to toluene. Scand J Work Environ Health 3: 135-143.

Cohr K-H, Stokholm J (1979). Exposure of humans to white spirit. IV. Concentrations in alveolar air and venous blood during house painting. Copenhagen: Arbejdstilsynet, Arbejdsmiljøinstituttet. Rapport nr. 3/1979: 27-51 (in Danish).

Perbellini L, Brugnone F, Mozzo P, Cocheo V, Caretta D (1984). Methyl ethyl ketone exposure in industrial workers. Uptake and kinetics. Int Arch Occup Environ Health 54: 73-81.

Sato A, Nakajima T (1979). Partition coefficients of some aromatic hydrocarbons and ketones in water, blood and oil. Br J Ind Med 36: 231-234.

Stokholm J, Cohr K-H (1979). Exposure of humans to white spirit. III. Concentrations in alveolar air and venous blood during experimental exposure. Copenhagen: Arbejdstilsynet, Arbejdsmiljøinstituttet. Rapport nr. 3/1979: 7-25 (in Danish).

Wigaeus E, Holm S, Åstrand, I (1981). Exposure to acetone. Uptake and elimination in man. Scand J Work Environ Health 7: 84-94.

Övrum P, Hultengren M, Lindqvist T (1978). Exposure to toluene in a photogravure printing plant. Scand J Work Environ Health 4: 237-245.

Åstrand I (1975). Uptake of solvents in the blood and tissues of man. Scand J Work Environ Health 1: 199-218.

Åstrand I, Ehrner-Samuel H, Kilbom Å, Övrum P (1972). Toluene exposure. I. Concentration in alveolar air and blood at rest and during exercise. Work Environ Health 9: 119-130.

Åstrand I, Engström J, Övrum P (1978). Exposure to xylene and ethylbenzene. I. Uptake, distribution and elimination in man. Scand J Work Environ Health 4: 185-194.

Åstrand I, Kilbom Å, Wahlberg I, Övrum P (1973). Methylchloroform exposure. I. Concentration in alveolar air and blood at rest and during exercise. Work Environ Health 10: 69-81.

Åstrand I, Kilbom Å, Övrum P (1975). Exposure to white spirit. I. Concentration in alveolar air and blood during rest and exercise. Scand J Work Environ Health 1: 15-30.

Åstrand I, Kilbom Å, Övrum P, Wahlberg I, Vesterberg O (1974). Exposure to styrene. I. Concentration in alveolar air and blood at rest and during exercise and metabolism. Work Environ Health 11: 69-85.
Åstrand I, Övrum P (1976). Exposure to trichloroethylene. I. Uptake and distribution in man. Scand J Work Environ Health 2: 199-211.
Åstrand I, Övrum P, Carlsson A (1975). Exposure to methylene chloride. I. Its concentration in alveolar air and blood during rest and exercise and its metabolism. Scand J Work Environ Health 1: 78-94.
Åstrand I, Övrum P, Lindqvist T, Hultengren M (1976). Exposure to butyl alcohol. Uptake and distribution in man. Scand J Work Environ Health 2: 165-175.

Uptake via the skin

Aitio A, Pekari K, Järvisalo J (1984). Skin absorption as a source of error in biological monitoring. Scand J Work Environ Health 10: 317-320.
Clendenning WE, Stoughton RB (1962). Importance of the aqueous/lipid partition coefficient for percutaneous absorption of weak electrolytes. J Invest Dermatol 39: 47-49.
Hansen CM (1982). The absorption of liquids into the skin. T 3-82 M+T. Scandinavian Paint and Printing Ink Research Institute.
Riihimäki V (1979). Percutaneous absorption of m-xylene from a mixture of m-xylene and isobutyl alcohol in man. Scand J Work Environ Health 5: 143-150.
Riihimäki V, Pfäffli P (1978). Percutaneous absorption of solvent vapors in man. Scand J Work Environ Health 4: 73-85.
Scheuplein RJ (1965). Mechanism of percutaneous adsorption. I. Routes of penetration and the influence of solubility. J Invest Dermatol, 45: 334-346.
Tsuruta H (1978). Percutaneous absorption of trichloroethylene in mice. Ind Health 16: 145-148.
Tsuruta H (1982). Percutaneous absorption of organic solvents. III. On the penetration rates of hydrophobic solvents through the excised rat skin. Ind Health 20: 335-345.
Treherne JE (1956). The permeability of skin to some non-electrolytes. J Physiol 133: 171-180.

Distribution and accumulation in the body.

Benignus VA, Muller KE, Graham JA, Barton CN (1984). Toluene levels in blood and brain of rats as a function of toluene level in inspired air. Environ Res 33: 39-46.

Carlsson A (1981). Distribution and elimination of C14-styrene in rat. Scand J Work Environ Health 7: 45-50.

Carlsson A (1981). Distribution and elimination of C14-xylene in rat. Scand J Work Environ Health 7: 51-55.

Pedersen LM, Larsen K, Cohr K-H (1984). Kinetics of white spirit in human fat and blood during short-term experimental exposure. Acta Pharmacol Toxicol 55: 308-316.

Carlsson A, Hultengren M (1975). Exposure to methylene chloride. III. Metabolism of C14-labelled methylene chloride in rat. Scand J Work Environ Health 1: 104-108.

Carlsson A, Ljungquist E (1982). Exposure to toluene. Concentration in subcutaneous adipose tissue. Scand J Work Environ Health 8: 56-62.

Engström J, Bjurström R (1977). Exposure to methylene chloride. Content in subcutaneous adipose tissue. Scand J Work Environ Health 3: 215-224.

Engström J, Bjurström R (1978). Exposure to xylene and ethylbenzene. II. Concentration in subcutaneous adipose tissue. Scand J Work Environ Health 4: 195-203.

Engström J, Bjurström R, Åstrand I, Övrum P (1978). Uptake, distribution and elimination of styrene in man. Scand J Work Environ Health 4: 315-323.

Engström J, Riihimäki V (1979). Distribution of m-xylene to subcutaneous adipose tissue in short-term experimental human exposure. Scand J Work Environ Health 5: 126-134.

Engström J, Åstrand I, Wigaeus E (1978). Exposure to styrene in a polymerization plant. Uptake in the organism and concentration in subcutaneous adipose tissue. Scand J Work Environ Health 4: 324-329.

Sullivan TM, Born GS, Carlson GP, Kessler WV (1983). The pharmacokinetics of inhaled chlorobenzene in the rat. Toxicol Appl Pharmacol 71: 194-203.

Wigaeus E, Löf A, Nordqvist M (1982). Distribution and elimination of 2-(14-C)-acetone in mice after inhalation exposure. Scand J Work Environ Health 8: 121-128.

Safety and Health Aspects of Organic Solvents, pages 61–71

METABOLISM AND EXCRETION OF ORGANIC SOLVENTS

Vesa Riihimäki

Department of Industrial Hygiene and Toxicology, Institute of Occupational Health, SF-00290 Helsinki, Finland

INTRODUCTION

Many organic solvents are extensively metabolized in the body. High metabolic activity implies that a significant share of the chemical entering the body (mainly through the lungs) is rapidly biotransformed and, consequently, effective uptake in the lungs continues throughout the exposure. In simple metabolic processes a high rate of production of metabolites is generally accompanied by a correspondingly rapid rate of metabolite excretion. In more complex systems, however, a certain delay in the production and excretion of end metabolites is often found. On the other hand, a low metabolic activity implies that only a small fraction of the chemical taken up is biotransformed. Regarding such compounds, when the concentrations in most tissues begin to reach equilibrium, uptake in the lungs decreases considerably. Since only small amounts of metabolites are produced, excretion mainly concerns the parent compound, notably via expired air.

Toxic action is intimately related to the kinetic fate of the substance, particularly metabolism. It can be stated that what the organism does to the chemical determines what the chemical does to the organism. For most organic solvents metabolism apparently counteracts toxicity by reducing tissue levels of the parent compound whose toxic action is, for example, based on its physical association with biological membranes. Furthermore, when the characteristically lipid soluble organic solvents are biotransformed to more polar compounds they are rendered unable to cross biological membranes and thus to reach potentially vulnerable sites. The metabolites are also more readily subjected to the processes of excretion.

The biotransformation of several solvents, e.g. benzene, carbon tetrachloride, n-hexane, methanol etc. is known to yield significant amounts of highly reactive metabolites (see Parkki, in this volume). For such compounds it can be expected that there is a direct correlation between the quantity of metabolism and the intensity of toxic effect. These solvents can be viewed as intrinsically toxic substances because their effects tend to be irreversible and to occur at relatively low dose levels whereas the general depressant effects common to all solvents are reversible and appear at high concentrations. This mechanistic distinction is certainly a very crude oversimplification of the true situation, however, because the nature of several known manifestations of solvent toxicity still remain to be clarified.

The metabolism of the common aromatic hydrocarbons, halogenated aliphatics, and a selected group of alkanes, alcohols and ketones have been under active investigation and their overall metabolic fates are rather well defined (for a review, see Toftgård and Gustafsson, 1980). On the other hand there are great gaps in knowledge regarding the metabolism and excretion of e.g. longer chain alkanes, higher alcohols and several ketones, ethers and esters.

SOLVENT METABOLIZING SYSTEMS

The metabolism of foreign compounds principally takes place in two phases. In phase I a polar group is introduced or unmasked by an oxidative, reductive or hydrolytic reaction. In phase II this reaction product is conjugated mainly with glucuronic acid, sulphate or glutathione and thus a highly polar compound is usually formed. Oxidative reactions such as aliphatic and aromatic hydroxylations and epoxidations are in most instances catalyzed by the cytochrome P-450 dependent monooxygenase system which primarily resides in the endoplasmic reticulum. Reductive reactions take place in both the microsomes and other cell fractions. Other phase I enzyme systems of significance for solvents are alcohol and aldehyde dehydrogenases and esterases which mainly occur in the soluble cell fraction.

The liver is doubtless the main metabolically active organ but relatively high activities of microsomal enzymes have occasionally been measured also in the intestinal mucosa, gonads, kidneys, lung and the skin. Among extrahepatic sites the lung and the skin are of special interest because they function as portals of entry to the body. Thus they are subjected to high concentrations before the compounds are distributed to different sites of action. Furthermore, the lung is maximally perfused. While the lung encompasses remarkable metabolizing activities for a selection of endogenous and xenobiotic substances (Bend et al., 1985) it is probably much inferior to the liver as regards organic solvent metabolism. Interestingly, there may be even qualitative differences of enzyme activity between the lungs and the liver, and the former is deficient in alcohol and aldehyde dehydrogenases which may result in an accumulation of potentially toxic aldehyde intermediates (Smith et al., 1982). The skin, on the other hand, is the largest organ of the body. At the present time there are insufficient data to estimate the roles and the significance of the different metabolizing tissues in intact organisms.

METABOLIC CAPACITY FOR SOLVENTS

Metabolic efficiency is often expressed by the fraction of the solvent metabolized in a balance study or by a measure of the metabolic rate such as, for instance, the retention in the lungs at steady-state or the calculated metabolic clearance. To illustrate the differences between compounds, table 1 presents a list of values of metabolic efficiency in man for a series of common organic solvents, modified and expanded from a presentation by Fiserova-Bergerova (1983). Most of the compounds appear to be metabolized extensively and some, notably 1,1,1-trichloroethane and tetrachloroethylene, to only a small extent. The fraction metabolized on the one hand, and the retention or clearance on the other, are not in perfect agreement and the values for tetrachloroethylene are conflicting. This points to a somewhat uncertain foundation of kinetic data apparently owing to variable methods of study. Nevertheless, xylene, styrene and trichloroethylene appear to be metabolized at very high rates. The clearance value obtained for trichloroethylene slightly exceeds the liver blood flow and, if it can be assumed that most metabolism occurs in the liver, it suggests that all trichloroethylene is extracted from the blood passing through

TABLE 1. Metabolic Efficiency for Selected Organic Solvents in Man

Compound	Fraction metabolized (%)	Pulmonary retention (%)	Metabolic clearance (l/min)
Toluene	80	50	0.51
Xylene	95	64	1.6
Styrene	95	64	1.6
Dichloromethane	70-95	31-51	0.74
1,1,1-Trichloroethane	5	18-25	
Trichloroethylene	75	35-60	1.85
Tetrachloroethylene	2	50	0.94
Acetone	80	45	
Methyl ethyl ketone	70-80	50	

this organ. During a short-term human exposure to styrene a clearance corresponding to the liver blood flow (about 1.6 l/min) was obtained (Wigaeus, 1983) after a reduction of 0.3 l/min for the continuing uptake by fat (Fiserova-Bergerova, 1983). The same figure emerged in a human exposure study on m-xylene, and we demonstrated that m-xylene concentration in the blood of inferior vena cava proximal to the portal veins was in the fasting state about 80 %, and after a meal about 64 %, of that in forearm venous blood (which can be regarded nearly comparable to arterial blood) indicating that extraction within the portal system was at least 0.6 - 0.8 (Riihimäki, 1979a).

Indeed it has been shown in the rat that at low concentrations of benzene, toluene, styrene, trichloroethylene, tetrachloroethylene, n-hexane and heptane, the metabolic extraction was large and that the blood flow to the metabolizing organ became a rate-limiting step (Andersen, 1981). When the metabolism is flow-limited the capacity of the liver can possibly be somewhat increased by a greater perfusion after a meal, and decreased by a lower absolute perfusion at heavy physical exertion (Rowell, 1974). At high tissue solvent concentrations enzymes can, however, become saturated. Thus for styrene, metabolic saturation occurs in the rat at environmental concentrations somewhere between 200 and 600 ppm (Ramsey and Young, 1978). When the metabolic capacity is exceeded the tissue concentrations of the parent compound will rise disproportionately, the formation and excretion of metabolites reach peak rates and consequently, elimination from the body will initially be delayed due to zero-order kinetics. Saturation of metabolism would thus tend to enhance the toxicity by the parent compound whereas that mediated by reactive or toxic metabolites would not increase. Saturation of the major biotransformation pathway could also lead to an activation of the minor but potentially more toxic pathway.

There is limited information regarding the maximal metabolic rates of solvents in man. Studies with m-xylene did not disclose any indications of metabolic saturation at environmental levels ranging from 100 to 280 ppm (Riihimäki, 1979a; Riihimäki et al., 1982). Therefore, concerning the efficiently metabolized solvents, saturation of biotransformation in the normal occupational circumstances is unlikely. Dichloromethane is an example of the opposite case. At low rates of uptake almost all dichloromethane is metabolized to carbon monoxide and carbon dioxide. At higher absorption rates the metabolized fraction decreases due to saturation of the metabolism; hence the pulmonary retention declines from about 50 % to 30 %. The capacity-limited production of carbon monoxide and formation of carboxyhemoglobin (COHb) at concentrations exceeding about 250 ppm was demonstrated in graded exposures. Exposure to 100 ppm of dichloroethane over 7.5 h at rest resulted in a COHb concentration of about 3.4 %, exposure to 200 ppm in 6.8 % (DiVincenzo and Kaplan, 1981) but exposure to 500 ppm only in 11.4 % (Stewart et al., 1973). Although only a very minor share of tetrachloroethylene is biotransformed to trichloroacetic acid and some other trichloro compounds, the excretion of total trichloro compounds increased exposure-dependently up till 100 ppm of tetrachloroethylene in the ambient air and then levelled-off which suggested metabolic saturation (Ohtsuki et al., 1983).

The metabolic handling of foreign compounds differs between individuals on the basis of genetic status, physiological variables and environmental influences (see Pelkonen, Døssing, in this volume). Physiological and environmental factors influence metabolism by enhancement or inhibition of enzyme activities. In the so-called induction the tissue has produced more enzyme to cope with an increased stress; inhibition is a condition where the enzyme is destroyed or, more commonly, unable to function at full capacity due to interference or competition by another compound. Induction or enhancement of metabolism by organic solvents themselves, e.g. aromatic hydrocarbons (Pyykkö, 1984; Toftgård, 1982; Andersen et al., 1984), and by other chemicals is not uncommon. However, for the efficiently metabolized solvents the occurrence of induction likely will not change their elimination since extraction in the metabolizing organ already in the normal state is virtually complete. Thus induction of microsomal enzymes with phenobarbital among volunteers who were then exposed to 92 ppm of m-xylene did not change the retention of xylene in the lungs or the excretion of methylhippuric acid in the urine (David et al., 1979). Only at high tissue concentrations that would normally saturate enzymes will the increased capacity become apparent. On the other hand, it has been demonstrated that enzyme inhibition and hence a reduced capacity to metabolize solvents may occur in a mixed solvent exposure (Engström et al., 1984) or under the influence of acute ethanol intake (Müller et al., 1975; Riihimäki et al., 1982), and cause profound alterations in kinetics.

PATHWAYS OF EXCRETION

Excretion of the unchanged solvent takes place mainly via the lungs and to a small and variable extent in the urine and other excreta such as sweat. Metabolites may be volatile and thus exhalable but mainly they are eliminated in the urine. Table 2 summarizes information concerning the elimination pathways and the specific compounds eliminated for some common organic solvents. There is a great diversity of excretion due to differences in solubility and metabolic characteristics between the compounds. A clear distinction can be made between the highly lipid soluble (e.g. aromatic) solvents and the water soluble compounds (e.g. ketones). The latter may occur unchanged in appreciable amounts in the urine (Wigaeus, 1983; Riihimäki, 1983) and saliva (Tomita and Nishimura, 1982) whereas the hydrophobic aromatics would not.

Furthermore, the time course of elimination is characteristically different between a lipid soluble solvent and a water soluble solvent. In the former case the solvent concentration in the blood exhibits several phases of elimination which mirror, initially, a rapid removal from the well-perfused internal organs and the muscles and then a slow

TABLE 2. Excretion Pathways of Selected Solvents in Man

Compound	Excretion in urine (%)		Excretion in exhaled air	
	after metabolism	unchanged	after metabolism	unchanged
Toluene	80 hippuric acid	0.06		7-21
Xylene	95 methyl-hippuric acid			3
Styrene	95 mandelic & phenyl-glyoxylic acid	traces		3
Dichloro-methane			70-95 carbon monoxide, carbon dioxide	5-30
1,1,1-tri-chloro-ethane	5 TCE*, TCA*			90
Tri-chloro-ethylene	75 TCE, TCA			10
Tetra-chloro-ethylene	2 TCA			95
Acetone	5 formate	1	20(?) carbon dioxide	20
Methyl ethyl ketone	10-20(?) 2,3-butane-diol	0.1		20

* TCE = trichloroethanol, TCA = trichloroacetic acid

elimination from adipose tissues. In the latter case, because water soluble compounds tend to be distributed rather evenly in the body water, only one apparent component of an intermediate elimination rate may be observed.

The same physiological variables which cause large interindividual differences in solvent uptake operate in the elimination which is essentially a reverse phenomenon (Fiserova-Bergerova et al., 1980). A large volume of distribution such as a large fat mass will prolong the elimination of principally lipid soluble solvents. Physical exercise during the elimination phase will enhance transport of the solvent from peripheral tissues to the sites of metabolism and excretion and thus speed up the elimination process.

EXCRETION MECHANISMS

Few studies have explored the excretion mechanisms for organic solvents and their metabolites. The kidneys are doubtlessly the organ of primary interest. Renal excretion is an integrated function of several processes: glomerular filtration, tubular secretion and reabsorption. Some organic acid-metabolites of solvents e.g. hippuric acid, methylhippuric acid and mandelic acid are efficiently removed by active tubular secretion (Kamienny et al., 1969; Riihimäki, 1979b). The capacity of their removal is probably clearly in excess of the maximal rates of their production. For sufficiently weak acids elimination may be modified by a subsequent reabsorption in the distal tubules at particularly low pH conditions, and urine flow rate might also influence reabsorption. The metabolites mentioned above will not, however, occur in an unionized form to any appreciable degree under the conditions normally met in kidney tubules and thus reabsorption presumably does not play a significant role. If a compound is tightly bound to plasma proteins it is less susceptible to renal excretion. This may explain why trichloroacetic acid, a metabolite of chlorinated hydrocarbon solvents, has an exceptionally long elimination half-time (Müller et al., 1975). Biliary excretion is another important route of elimination for foreign compounds. Excretion of solvents or metabolites in the bile of man is largely unexplored.

In experimental human exposures to solvents, performed

under well-controlled conditions, a considerable variation is typically found in the amounts of metabolites excreted in the urine. For a compound that is mainly transformed to metabolites this appears to depend primarily on the different quantities of the substance taken up in the lungs owing to intra- and interindividual variations of pulmonary ventilation. Even greater variation of urinary metabolites may, however, concern compounds with low metabolism, e.g. 1,1,1-trichloroethane (Nolan et al., 1984) indicating that the interindividual differences in the biotransformation and excretion may play a great role.

REFERENCES

Andersen ME (1981). Pharmacokinetics of inhaled gases and vapors. Neurobehav Toxicol Teratol 3: 383-389.

Andersen ME, Gargas ML, Ramsey JC (1984). Inhalation pharmacokinetics: Evaluating systemic extraction, total in vivo metabolism, and the time course of enzyme induction for inhaled styrene in rats based on arterial blood: inhaled air concentration ratios. Toxicol Appl Pharmacol 73: 176-187.

Bend JR, Serabjit-Singh CJ, Philpot RM (1985). The pulmonary uptake, accumulation, and metabolism of xenobiotics. Ann Rev Pharmacol Toxicol 25: 97-125.

David A, Flek J, Frantik E, Gut I, Sedivec V (1979). Influence of phenobarbital on xylene metabolism in man and rats. Int Arch Occup Environ Health 44: 117-125.

DiVincenzo GD, Kaplan CJ (1981). Uptake, metabolism and elimination of methylene chloride vapor by humans. Toxicol Appl Pharmacol 59: 130-140.

Engström K, Riihimäki V, Laine A (1984). Urinary disposition of ethylbenzene and m-xylene in man following separate and combined exposure. Int Arch Occup Environ Health 54: 355-363.

Fiserova-Bergerova V (1983). Modeling of metabolism and excretion in vivo. In Fiserova-Bergerova V (ed): "Modeling of Inhalation Exposure to Vapors: Uptake, Distribution and Elimination," vol I, Boca Raton, Florida: CRC Press, pp 102-128.

Fiserova-Bergerova V, Vlach J, Cassady JC (1980). Predictable "individual differences" in uptake and excretion of gases and lipid soluble vapours simulation study. Br J Ind Med 37: 42-49.

Kamienny FM, Barr M, Nagwekar JB (1969). Mutual inhibitory effect of (-) mandelic acid and certain sulfonamides on the kinetics of their urinary excretion in humans. J Pharm Sci 58: 1318-1324.
Müller G, Spasskowski M, Henschler D (1975). Metabolism of trichloroethylene in man. III. Interaction of trichloroethylene and ethanol. Arch Toxicol 33: 173-189.
Nolan RJ, Freshour NL, Ride DL, McCarthy LP, Saunders JH (1984). Kinetics and metabolism of inhaled methylchloroform (1,1,1-trichloroethane) in male volunteers. Fund Appl Toxicol 4: 654-662.
Ohtsuki T, Sato K, Koizumi A, Kumai M, Ikeda M (1983). Limited capacity of humans to metabolize tetrachloroethylene. Int Arch Occup Environ Health 51: 381-390.
Pyykkö K (1984). "Toluene and Regulation of Cytochrome P-450 in the Rat." Tampere: University of Tampere.
Ramsey JC, Young JD (1978). Pharmacokinetics of inhaled styrene in rats and humans. Scand J Work Environ Health 4:suppl 2, 84-91.
Riihimäki V (1979). "Studies on the Pharmacokinetics of m-Xylene in Man." Helsinki: Institute of Occupational Health and Tampere: University of Tampere.
Riihimäki V (1979b). Conjugation and urinary excretion of toluene and m-xylene metabolites in a man. Scand J Work Environ Health 5: 135-142.
Riihimäki V (1983). Methyl ethyl ketone (MEK). Nordic expert group for documentation of occupational exposure limits. Arbete och hälsa 25. Stockholm: Arbetarskyddsverket (in Swedish).
Riihimäki V, Savolainen K, Pfäffli P, Pekari K, Sippel HW, Laine A (1982). Metabolic interaction between m-xylene and ethanol. Arch Toxicol 45: 253-263.
Rowell LB (1974). Human cardiovascular adjustments to exercise and thermal stress. Physiol Rev 54: 75-159.
Smith BR, Plummer JC, Wolf CR, Philpot RM, Bend JR (1982). p-Xylene metabolism by rabbit lung and liver and its relationship to the selective destruction of pulmonary cytochrome P-450. J Pharmacol Exp Ther 223: 736-742.
Stewart RD, Forster HV, Hake CL, Lebrun AJ, Peterson JE (1973). Human responses to controlled exposures of methylene chloride vapor. Report No NIOSH-NCOW-Envm-MC-73-7. Milwaukee: The Medical College of Wisconsin.
Toftgård R (1982). "Xylene: an Inducer of Hepatic Cytochrome P-450 and Xenobiotic Biotransformation in the Rat." Stockholm: Karolinska Institutet.

Toftgård R, Gustafsson J-Å (1980). Biotransformation of organic solvents. A review. Scand J Work Environ Health 6: 1-18.
Tomita M, Nishimura M (1982). Using saliva to estimate human exposure to organic solvents. Bull Tokyo Dent Coll 23: 175-188.
Wigaeus E (1983). Kinetics of acetone and styrene in inhalation exposure. Arbete och hälsa 23. Stockholm: Arbetarskyddsverket.

Safety and Health Aspects of Organic Solvents, pages 73–87

SIMULATION MODELS FOR ORGANIC SOLVENTS

Pierre O. Droz

Institute of Occupational Medicine and Industrial Hygiene, University of Lausanne
rte de la Clochatte - 1052 Le Mont - Switzerland

INTRODUCTION

The toxicity of organic solvents is determined by a two-step process : (1) the transfer of the chemical from the environment to the site of action. (2) the interaction of the solvent or the active species with the receptor leading to acute or chronic effects. Toxicokinetics refers to the study of the first step : kinetics of the apperance and disapperance of the solvent and its metabolites in the tissues of toxicological interest.

The behavior of organic solvents in the body can be studied by 3 main approches :

- experimental studies, where volunteers are exposed in well-controlled conditions and the kinetics of the solvents or the metabolites are studied in various biological media (expired air, urine, blood, fatty tissues),
- field studies, where workers exposed under field conditions are investigated using a limited number of biological samples in order to deduce kinetic properties of the solvent and the metabolites,
- theoretical studies, using different kinds of models to allow the simulation of various exposure situations and individual characteristics.

Each of these three approaches has its own advantages, but also its own limitations. For example, experimental studies give very good toxicokinetic data describing isolated exposures but they are often of little help in the understanding of field situations if a mathematical model is not constructed from the experimental data. With regard to field investigations on the other hand, there are so many factors (environmental variability, interindividual differences) arising simultaneously that very often the results obtained are difficult to interpret.

The use of mathematical models can be considered as complementary to both the experimental and field approaches. On the one hand they can be used to extrapolate experimental results to field situations, and on the other hand they allow a better understanding and prediction of the multifactor field situations.

FUNDAMENTALS OF COMPARTMENTAL MODELS

Mathematical models have now been used for many years in the description of the fate of chemicals in the body. Different types of models can be distinguished depending on the basic hypotheses and the level of sophistication.
Empirical models. They consist of empirical mathematical formulas (usually exponentials) the parameters of which are found by best fitting to experimental results.
Toxicokinetics models. In this case the mathematical formulas are based on a simple model of the distribution and metabolism (one, two-compartments) but the parameters are still found by best fitting to the experimental results.
Simulation models. In this type of model the mathematical relationships and their variables have a physiological and metabolic meaning and thus are not found by fitting the experimental results.

All three types of models are very useful tools for the description of the behavior of solvents in the body. Nevertheless it must be recognized that simulation models are much more flexible than the first two types. They allow notably the study of new situations, of the effect of physiological (body build, physical workload) and metabolic factors on the transport of the solvent to the active site.

Only simulation models will thus be considered here. First developed by anaesthesiologists (Papper and Kitz 1963) they have then been applied to industrial toxicology to describe the behavior of organic volatiles in workers occupationally exposed (Fiserova et al. 1974, Gubéran and Fernandez 1974). About 20 industrial solvents have now been studied using simulation models, but only a few have been thoroughly compared with human experimental data for validation.

Description of a Simulation Model

The simulation of the behavior of organic solvents can be separated into two parts : the simulations of the solvent itself and the description of the formation and elimination of the metabolites. To simulate the solvent's behavior, the body is considered to consist of five homogeneous compartments containing tissues of similar perfusion/volume ratio (Fiserova 1983) :

1. A pulmonary compartment, responsible for exchanges with the gaseous environment.
2. A group of vessel-rich tissues (VRG) containing the heart, the brain, the kidneys, etc.
3. A group of low perfused tissues (MG) representing mainly the muscles and the skin.
4. A group of poorly perfused fatty tissues(FG).
5. A tissue responsible for the metabolic breakdown of the solvent (liver).

The description of the exchanges between air, tissues and blood is done on the basis of the following main hypotheses :

1. The solvent diffuses freely and instantaneously through the entire surface of the capillaries and alveolar walls.
2. The solvent in venous blood is in equilibrium with that dissolved in the corresponding tissue.
3. The concentration in arterial blood is in equilibrium with that of the alveolar air.

Figure 1 gives a schematic representation of the body for the distribution of the solvents.

The description of the metabolic phenomena is usually a more complex process and is variable from one solvent to the other. Figure 2 shows a basic schema which can be used to describe, in a simple way, the metabolism of some of the most widely used solvents (Fernandez et al. 1977, Droz and Guillemin 1983, Caperos et al. 1982).

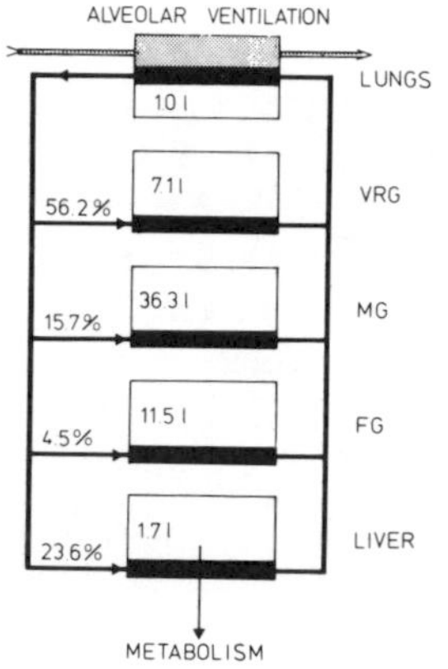

Figure 1. Schematic presentation of the compartmental model. The blood perfusion (%) and tissues volumes (l) are indicated for a standard man.

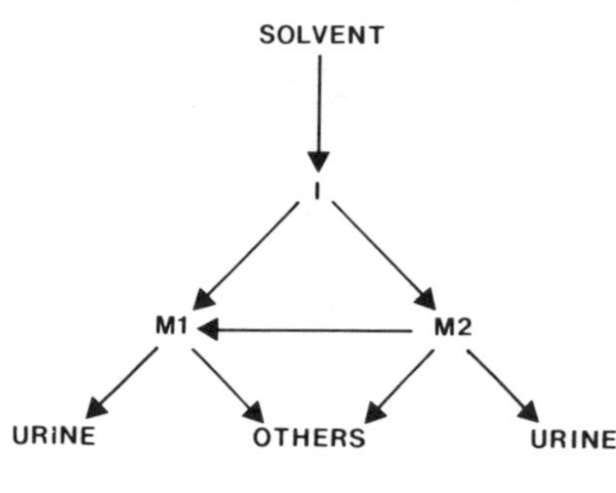

Figure 2. Schematic presentation of the biotransformation and excretion pathways used to simulate the metabolism of solvents.

The solvent is supposed to be transformed instantaneously into an unstable intermediate, which can give rise potentially to 2 metabolites. The latter can then transform into each other and/or be excreted into urine or other excreta. In order to describe the metabolism with the least experimental data as possible, the following assumptions are generally made :

1. All the reactions are considered to be of the first order type in the concentration ranges occuring during occupational exposures.
2. The transformation into metabolite 1 or 2 is considered as being much faster than the metabolites' kinetic behavior.
3. The distribution of each metabolite in the body takes place in one single volume.

It is to be noted that, as long as only excretion rates are needed, the volume of distribution can be omitted.

Physiological, Physicochemical and Metabolic Parameters

Three types of data are needed to completely solve the model described above :

1. Physiological : the tissues and blood volumes and the blood perfusions.
2. Physicochemical : the solubility or partition coefficient of the solvents in the various tissues and in blood.
3. Metabolic : clearance and excretion rates of the metabolites.

The physiological characteristics of the compartments described before can be found in medical tables and references (Fiserova 1983, Eger 1974). Table 1 presents the tissue volumes and blood perfusions used in the model for a 70 kg standard man.

The physico-chemical properties of the solvents, their tissue-gas partition coefficients, have been reviewed on several occasions (Fiserova 1983, Stewart et al. 1973). Missing data can be estimated through approximations using water and oil partition coefficients (Droz 1978). Table 2 summarizes some partition coefficients for the most widely used industrial solvents.

The metabolic clearance, or biotransformation rate of the solvent in the organism , can be estimated using different techniques based on solvent retention or metabolite output (Fiserova 1983, Droz and Fernandez 1977). The kinetic behavior of the metabolites can be reassessed either from human or animal studies.

VALIDATION OF THE MODEL

The development of compartmental models describing the absorption, metabolism and excretion of solvents is based on numerous hypotheses. The first logical step, before using these models, is thus to validate the hypotheses by comparison of predicted results with data obtained in experimental

TABLE 1. Physiological parameters for a standard man (70kg)

Compartment	Volume [1]	Perfusion [1/min]
Lungs air	2.85	6.0*
tissue	1.0	6.3
VRG	7.1	3.5
MG	36.3	1.0
FG	11.5	0.28
Liver	1.7	1.5

* Alveolar ventilation BTPS

TABLE 2. Gas tissues or liquids partition coefficients

Solvent	Blood	Oil	Water	VRG[1]	MG[2]	FG[3]
Benzene	7	498	2.8	15	10	350
Toluene	16	1460	2.5	30	23	1030
m-Xylene	34	4321	2.2	80	60	3030
Styrene	59	5838	4.9	150	84	4100
Methylene chloride	8	157	6.5	17	11	260
Chloroform	11	424	3.8	16	11	300
Trichloroethylene	9	763	1.5	20	19	600
Methylchloroform	4	373	0.9	9	6	373
Tetrachloroethylene	14	2072	0.9	45	29	2070

[1]Estimated with $\lambda_{heart}/_{H_2O}=0.0172\lambda_{oil}/\lambda_{H_2O}+2.3$

[2]Calculated from $\lambda_{MG}/\lambda_{H_2O}=0.0133\lambda_{oil}/\lambda_{H_2O}+1.362$

[3]Calculated from $\lambda_{FG}=0.7\cdot\lambda_{oil}+0.3\lambda_{blood}$

or field studies.

Only a few solvents have been fully validated up till now. Nevertheless, the confirmation obtained in a few cases allows us to generalize to other untested solvents exhibiting the same general properties.

Mainly the levels of solvents or metabolites in two principal biological media can be used in the validation : solvent in expired air, metabolites in urine (eventually in blood). Figure 3 shows the results obtained in such a comparison for trichloroethylene in alveolar air (Fernandez et al. 1977). The agreement between the theoretical curve predicted by the model and the experimental points is very satisfactory and indicates that the hypotheses concerning the solvent's distribution are realistic. Similar results have been shown for styrene (Droz and Guillemin 1983), methylchloroform (Caperos et al. 1982), methylene chloride (Hake 1979), tetrachloroethylene (Guberan and Fernandez 1974) and benzene (Fiserova et al. 1974). Figure 4 presents the results obtained for the urinary excretion of mandelic acid (MA) after experimental exposure to styrene (Droz and Guillemin 1983). Again the agreement is good indicating that the simplified metabolic pathways assumed are a satisfactory model. Good comparative results have also been published for trichloroethylene and methylchloroform. The validation can be extended to sub-acute situations with data obtained during one week of repeated exposure. Such a comparison is shown in figure 5 for trichloroethylene (Fernandez et al. 1977). The agreement is very good indicating that the simple model used to simulate the metabolism is valid also under repeated conditions. Such models can therefore be used to simulate industrial type repeated exposures. Figure 6 presents the comparative results obtained while simulating a whole week's exposure for two styrene workers. The results obtained for uptake, MA and phenylglyoxylic acid (PGA) agree well with the values measured (Droz et al. 1984). Although complete validation is still limited to a small number of chemicals, the basic hypotheses are probably realistic and the results obtained for new compounds can be used for a first approximation.

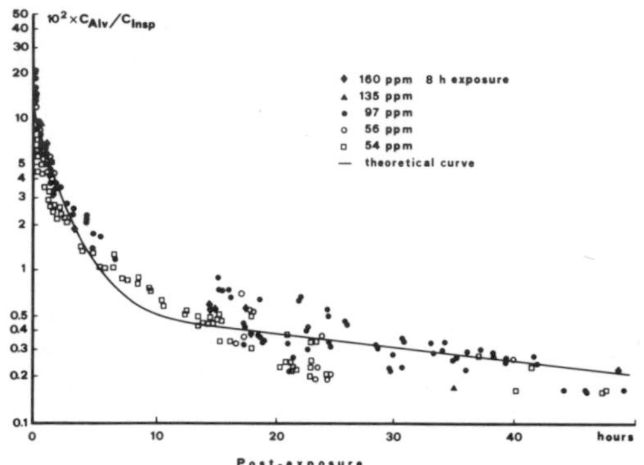

Figure 3. Alveolar concentrations of trichloroethylene after 8-hour exposure. Comparison between model and experimental data (Fernandez et al. 1977, reproduced with the permission of Brit. J. Ind. Med.).

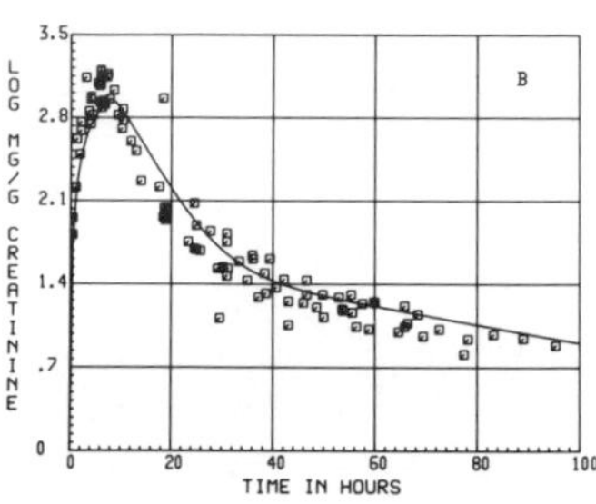

Figure 4. Urinary excretion of mandelic acid after 8 hours' exposure at 100 ppm. Comparison between model and experimental data (Droz and Guillemin 1983, reproduced with the permission of Int. Arch. Occup. Environ. Health).

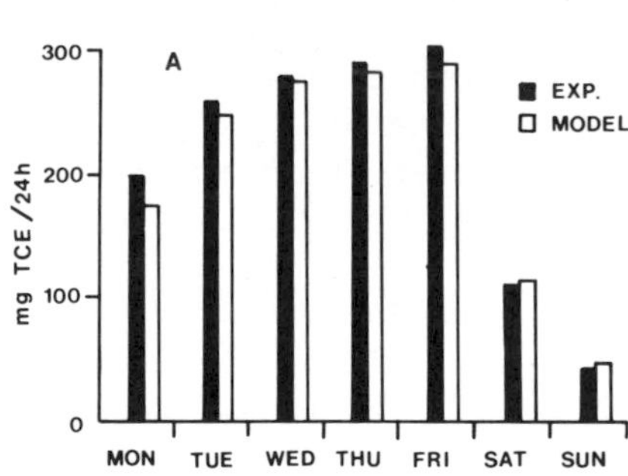

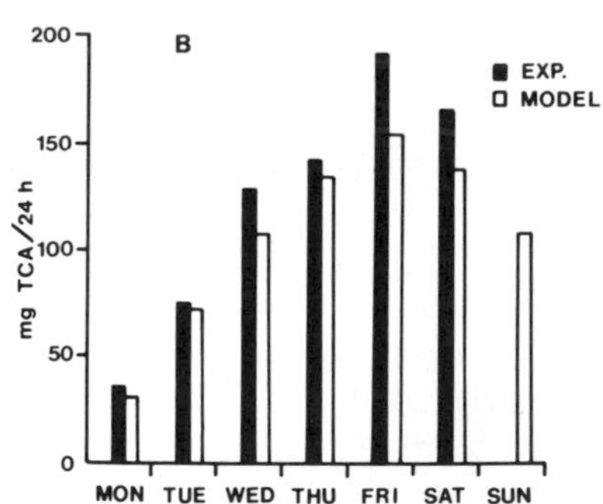

Figure 5. Daily urinary excretion of trichloroethanol and trichloroacetic acid during a week's exposure. Comparison of model and experimental results.

APPLICATION OF COMPARTMENTAL MODELS

Once validated by comparison with experimental data, simulation models can in principle be used to study many aspects refering to the toxicokinetics of solvents. Only a few examples of the possibilities will be given here :

- simulation of tissue distribution and body burden,
- establishment of biological indicators of exposure,
- evaluation of the effect of individual and environmental factors,
- interaction of solvents with other chemicals such as alcohol.

The main use up till now has been in the biological monitoring field probably because kinetic data were absolutely needed in order to interpret biological indicator data. An example of the simulation of tissue distribution is shown in figure 7 for trichloroethylene during a 100 ppm 8-hour exposure (two 4-hour exposure periods with a 1-hour break) on Thursday of a steady-state week. It clearly shows the different behaviors of a slow compartment (FG), an intermediate compartment (MG) and the fast compartments (VRG, blood) during stepwise changes in the exposure concentration. Environmental variability will thus affect tissue concentrations differently depending on the kinetic properties of the tissues of interest. Figure 8 shows an example of the body burden of trichloroethylene, trichloroacetic acid and trichloroethanol during a one week of fluctuating exposure. In this case we can see that chemicals (solvents and metabolites) having different kinetic properties will behave in completely different ways in the organism.

Simulation models have been used many times in the development and the understanding of biological monitoring. They can serve both in the setting up of biological limit values by simulation of repeated exposures and in the study of the effect of various factors (environmental and individual variability, timing) on those limits. Tables 3 gives examples of biological limit values for breath analysis established using a simulation model, and the corresponding results obtained from experimental of field studies. In most cases the agreement between the two sets of values is quite satisfactory.

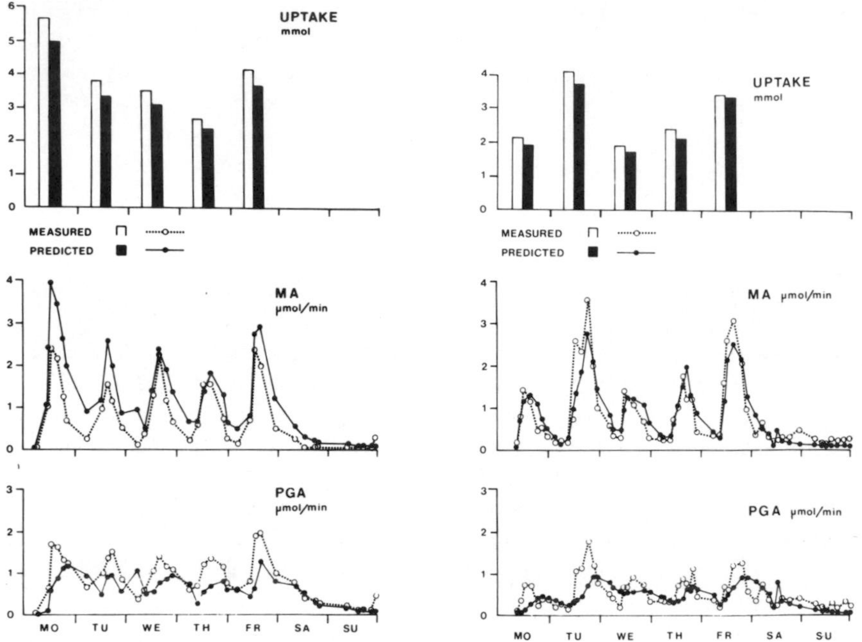

Figure 6. Styrene uptake and the urinam excretion of mandelic and phenylglyoxylic acids for two workers during a full week. Comparison between model and measured results.

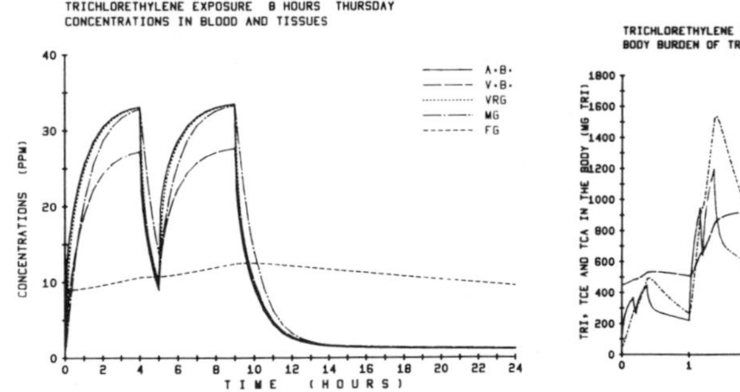

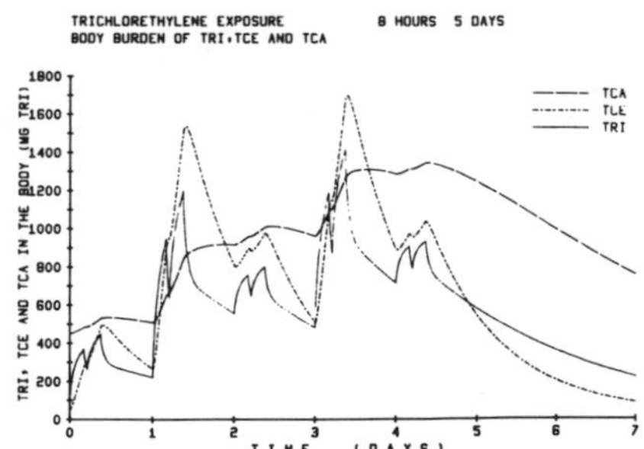

Figure 7. Blood and tissue concentrations of trichloroethylene during 8 hours (4/1/4) exposure at 100 ppm on a Thursday, steady-state week.

Figure 8. Body burdens of TRI, TCE and TCA during one week of fluctuating exposure (100, 300, 100, 300, 100 ppm).

The estimation of the effect of individual factors such as body build, pulmonary ventilation, metabolic characteristics can easily be undertaken with the aid of simulation models. Several papers (Fiserova et al. 1980, Fiserova 1985, Droz 1985) have already been devoted to this aspect with the aim of improving biological monitoring. As an example Table 4. shows the estimated bias produced by various factors on breath analysis results, for samples taken at the end of the exposure and 15 hours later (Droz 1985).

Finally, simulation models can be used and will be used in the future to obtain better understanding of the interaction between chemicals. As an example, the case of alcohol-solvent interaction can be approached theoretically. With styrene, for example, it has been shown experimentally that alcohol inhibits the oxidation of styrene glycol (an intermediate in styrene metabolism) into mandelic acid (Wilson 1985). This effect can be simulated using a very simple yes/no model where the interaction is either present and total or absent. Figure 9 shows the results obtained during 6-hour exposure with intake at mid-exposure of 0.25 and 0.5 g/kg body weight of alcohol.

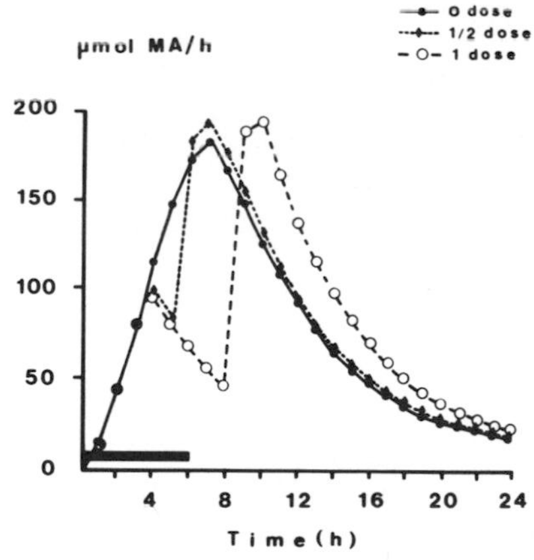

Figure 9. Mandelic acid urinary excretion during and after 6-h expoure at 50 ppm styrene and with 0.25 and 0.5 g alcohol/kg body weight.

The effect of the highest dose of alcohol on mandelic acid is remarkable and it is similar to that obtained under experimental conditions. This type of a simple model can

TABLE 3 Comparison between experimental and predicted breath concentrations sampled after 0.5 and 15 hours of postexpos

SOLVENT (TLV)		Model prediction [ppm]	Experimental [ppm]
Methylchloroform	0.5 h	120	273
(350 ppm)	15 h	16	17
Tetrachloroethylene-	0.5 h	25	23
ne (50 ppm)	15 h	7	8
Trichloroethylene	0.5 h	7	3-4
(50 ppm)	15 h	0.6	0.3-0.4
Benzene (10 ppm)	0.5 h	1.5	
	15 h	0.2	0.12
Styrene (50 ppm)	0.5 h	2	
	15 h	0.3	0.06

TABLE 4 Average bias on breath concentrations produced by various factors.

FACTOR	BIAS [%]	
sampling time	0.5 h	15 h
Repetition of exposure	10	60
Interday exposure variations	80	5
Intraday exposure variations	5	25
Physical workload	15-50	5-30
Metabolic rate	1-70	2-100
Body build	5	55

be used to simulate real life conditions to study if normal drinking habits can significantly alter the metabolic and kinetic behavior of solvents.

CONCLUSIONS

Simulation techniques using compartmental models have now been in use for many years and their development and validation is advanced enough to allow their use in the discussion of solvent toxicity. They can greatly improve our understanding of the kinetic behavior of solvents and their biotransformation products not only under simple steady conditions but also during complex fluctuating exposure situations.

Their use has mainly been limited up till now to the understanding of biological monitoring results and strategy. In the future they could be more widely applied to toxicological investigations.

REFERENCES

Caperos JR, Droz PO, Hake CL, Humbert BE, Jacot-Guillarmot A (1982). 1,1,1-trichloroethane exposure. Biological monitoring by breath and urine analyses. Int. Arch. Occup. Environ. Health 49: 293-303.

Droz PO, Fernandez JG (1977). Effect of physical workload on retention and metabolism of inhaled organic solvents. A comparative theoretical approach and its application with regards to exposure monitoring. Int. Arch. Occup. Environ. Health 38: 231-246.

Droz PO (1978). Contribution à la recherche d'indices biologiques d'exposition aux solvants: détermination de leurs coefficients de partage et étude de leurs comportements dans l'organisme à l'aide de modèles de simulation. Neuchâtel University, Switzerland [Thesis].

Droz PO, Guillemin MP (1983). Human styrene exposure. V. Development of a model for biological monitoring. Int. Arch. Occup. Environ. Health 53: 19-36.

Droz PO, Guillemin MP (in press). Occupational exposure monitoring using breath analysis. J. Occup. Med.

Droz PO (in press). The use of simulation models for setting BEIs for organic solvents. Annals of the American Conference of Governmental Industrial Hygienists.

Droz PO, Sollenberg J, Bjurstroem R, Versterberg O, (1984). Occupational styrene exposure : monitoring of uptake and analysis of metabolites in urine. II. Comparison of a computer model with analysis of data. Presented at the International Conference on Organic Solvents Toxicity, Stockholm (S).

Eger EI ed. (1974). Anesthetic uptake and action. The Williams & Wilkins Company, Baltimore.

Fernandez JG, Droz PO, Humbert JE, Caperos JR (1977). Trichloroethylene exposure: simulation of uptake, excretion and metabolism using a mathematical model. Br. J. Ind. Med. 34: 43-55.

Fiserova-Bergerova V, Vlach J, Singhal K (1974). Simulation and prediction of uptake, distribution and extraction of organic solvents. Br. J. Ind. Med. 31: 45-52.

Fiserova-Bergerova V, Vlach J, Cassady JC (1980). Predictable "individual differences" in uptake and excretion of gases and lipid soluble vapours simulation study. Br. J. Ind. Med. 37: 42-49.

Fiserova-Bergerova V ed. (1983). Modelling of inhalation exposure to vapors: uptake, distribution and elimination. CRC Press. Boca Raton F 1, vol. 1 and 2.

Fiserova-Bergerova V (in press). Simulation model as tool for adjustment of biological exposure indices to exposure conditions. Dillon K, Ho MH ed.: Biological monitoring of exposure to chemicals. vol 1, Organic compounds. John Wiley & Sons, New-York.

Gubéran E, Fernandez JG (1974). Control of industrial exposure to tetrachloroethylene by measuring alveolar concentrations: theoretical approach using a mathematical model. Br. J. Ind. Med. 31: 159-167.

Hake CL (1979). Simulation studies of blood carboxyhemoglobine levels associated with inhalation exposure to methylene chloride. Presented at the 18th annual meeting of the Society of Toxicology, New Orleans.

Papper EM, Kitz RJ ed. (1963). Uptake and distribution of anesthetic agents. McGraw Hill, New-York.

Stewart A, Allott PR, Cowles AL, Mapleson WW (1973). Solubility coefficients for inhaled anaesthetics for water, oil and biological media. Br. J. Anaesth. 45: 282-293.

Wilson HK, Robertson SM, Waldron HA, Gompertz D (1983). Effect of alcohol on the kinetics of mandelic acid excretion in volunteers exposed to styrene vapour. Br. Ind. Med. 40: 75-80.

Safety and Health Aspects of Organic Solvents, pages 89–96

BIOTRANSFORMATION REACTIONS AND ACTIVE METABOLITES

M.G. Parkki,

Department of Physiology, University of Turku, Finland

Conventional view of toxicity

Because most of the classical poisons act unchanged, the role of the metabolism in toxic effects of various chemicals stayed quite long without closer attention. Binding of the effector to a certain "receptor" eg. in a nerve synapse or at the active centre of an enzyme was known as a mechanism of toxicity since the days of Paul Erlich at the beginning of this century. It was also easy to understand cell destruction by caustic compounds or lipid solvents. These mechanisms gave however no explanation to the toxicity of many apparently harmless compounds. Moreover, when toxicology began to concentrate on the chronic effects and adopted new fields like carcinogenesis mutageneses and teratogenesis, new theories were needed.

The concept of metabolism and reactive intermediates

The role of metabolism in the activation of a foreign compound was for the first time shown conclusively with pronotosil, the first sulphonamide antibiotic in 1935. A second step was the realization, that the metaboites can also be reactive. It was observed, that azo-dyes were bound to tissues in vivo (Miller and Miller 1947). This did not happen in vitro indicating the necessity of metabolism of the parent compound. The concept of reactive intermediates formed during metabolism was rapidly applied as a mechanism for carcinogenicity, mutagenicity, teratogenicity, cell toxicity and for allergic responses (Jollow et al. 1977, Snyder et al. 1982).

The role of metabolism

Many of the foreign compounds, both natural and manmade, are lipid soluble. In this form they cannot be excreted from the organism. Accumulation of these compounds into the cells would soon impair their function. This is why biotransformation must be undertaken. In many cases the molecules posses stable structures. The transformation of these molecules needs violent treatment and the high energy required is temperately transferred to the molecule giving rise to reactive species. If the enzymes fail to handle these intermediates in a controlled way, they escape and cause cell damage by binding irreversibly to important cell constituents or start a harmful chain reaction in the membrane lipids called lipid peroxidation.

Various forms of reactive intermediates and enzymes involved

Most of the reactive intermediates are formed in the so-called first phase reactions of drug metabolism, which commonly are oxidative in nature. The most important first-phase enzymes are the cytochrome P-450 mediated monooxygenases locating in the endoplasmic reticulum (microsomal monooxygenases). Thus these enzymes are also most often involved in the production of reactive intermediates from various xenobiotics. In the normal monooxygenation reaction one molecular oxygen is consumed and two electrons are mediated by cytochrome P-450 resulting in an oxidative alteration of the substrate and the formation of one water molecule. Epoxides and N-hydroxycompounds are formed by this kind of real monooxygenation reaction. Aliphatic epoxides may rearrange to further reactive forms, eg. aldehydes. The N-hydroxy compounds form radicals after further rearrangements. In addition to the real monooxygenations the cytochrome P-450 mediated monooxygenases may in a way, fail to complete the reaction after donation of the first eletron. Thus actually a reductive reaction produces a radical. Radicals are a wide group of ractive intermediates with odd number of electrones. Also many other red-ox enzymes produce radicals from various compounds. One of these enzymes is the electron donator of cytochrome P-450: the cytochrome P-450 reductase, which is a flavoprotein. For references, see Jollow et al. 1977, Hodgson et al. 1979, McBrien and Slater 1982, Snyder et al. 1982.

An important group of radicals are the various oxygen radicals, which represent different reduced forms of the oxygen molecule. These can also be produced by cytochrome P-450, cytochrome P-450 reductase and other red-ox enzymes. Some radicals of foreign compounds eg. quinones and antibiotic cytostatics are able to cycle eg. with cytochrome P-450 reductase to produce catalytically large amounts of reactive oxygen (Pryor 1982, McBrien and Slater 1982).

Although the second phase reactions are traditionally regarded as real detoxication reactions many reactive intermediates are found among conjugates. The ultimate reactive species of carcinogenic azo-dyes are sulphate and glucuronide conjugates (Miller and Miller 1982).

The mechanisms of toxicity by the reactive intermediates

Covalent binding

The reactive intermediates are able to react with nucleophilic groups of cell constituents forming stable covalent bonds. If these targets are essential for the integrity of the cell function, eg. the genetic code, mutagenic, carcinogenic, teratogenic, and cell toxic responses may ensue, for references, see Jollow et al. 1977, Snyder et al. 1982.

Lipid peroxidation

Lipid peroxidation is a chain-reaction affecting the lipid membranes of the cell. In addition to membrane destruction, covalent interactions and DNA-damage may be involved. Lipid peroxidation is started by organic and oxygen radicals. The reaction leads to acute cell and tissue damage, but it is also linked to eg. ageing and carcinogenesis and related processes. For references, see Hodgson et al. 1979, Pryor 1982, McBrien and Slater 1982.

Systems protecting against reactive intermediates

Two enzymes, epoxide hydrolase and glutathione S-transferase are capable of inactivating epoxides. The former adds water to the reactive epoxide giving rise to rather nontoxic dihydrodiols (Lu and Miwa 1980). Glutathione S-transferase catalyzes the conjugation of epoxide with glutathione acceptor (Jakoby and Habig 1980). These reactions of epoxides can also proceed nonenzymatically.

The reactive oxygen species can be inactivated enzymatically by superoxide dismutase and glutathione peroxidase. For references, see Hodgson et al. 1979, Jakoby 1980, Pryor 1982.

Radicals of organic compounds cannot be inactivated enzymatically and the reactions always take place randomly. The reaction can occur safely with glutathione. There exist also other radical scavangers like vitamin E, which are effective at least in vitro. The covalent binding of reactive intermediates with noncritical targets like cytochrome P-450. ligandin, metallothionein and many other proteins can be regarded as "detoxication."

Reactive intermediates of organic solvents

Organic solvents produce various forms of toxicity both acutely and chronically. Both acute and chronic effects are seen in the function of the nervous system. Of the other organs affected especially the liver is often involved. Also in the connection with many solvents carcinogenic, mutagenic and related forms of toxicity have been observed. Only some of the acute effects can be explained by the action of the compound itself. In most cases the reactive forms arising during metabolism are associated with the toxic effects.

Frequently the metabolism of a compound involves several reactive intermediates and it is not clear, which of them is responsible for the toxic effects. It may actually happen that different forms of toxicity are mediated by different intermediates. On the other hand several intermediates and mechanisms may be operating at the same time.

EXAMPLES:

Benzene

The rather unique property of benzene is its mayelotoxicity. It causes disturbances in leuco- and eryhropoiesis of varying severity up to agranulo sytosis and leucemia. At high doses effects in other organs like the nervous system and the liver arise. The metabolic pathways of benzene are seen in figure 1. The first step is the epoxidation by cytochrome P-450 mediated microsomal monooxygenases. Benzene oxide can be isomerized to phenol, hydrated to dihydrodiol or conjugated with glutathione. Phenol is believed to form hydroquinone, semiquinone and quinone as as a result of a second oxidation by P-450. The dihydrodiol may be reduced to pyrocatechol, which in turn after secondary oxidation forms the quinones. Benzene is the only aromatic solvent believed to undergo a ring scission to produce mucoaldehyde and muconic acid. Of the metabolites of benzene, the epoxide, cathecols, semiquinones, quinones and muconaldehyde are reactive species being able eg. to bind covalently, for references, see Snyder et al. 1982. At the present time the ultimate myelotoxin of benzene is not known with certainty.

Styrene

In man the most important form of toxicity is affection of the neural system. Also indices of genotoxic effects have been demonstrated. Suspicion about hematopoietic and lymphatic malignancies in styrene workers exist. Experimentally, high doses produce also hepato- and renotoxic effects.

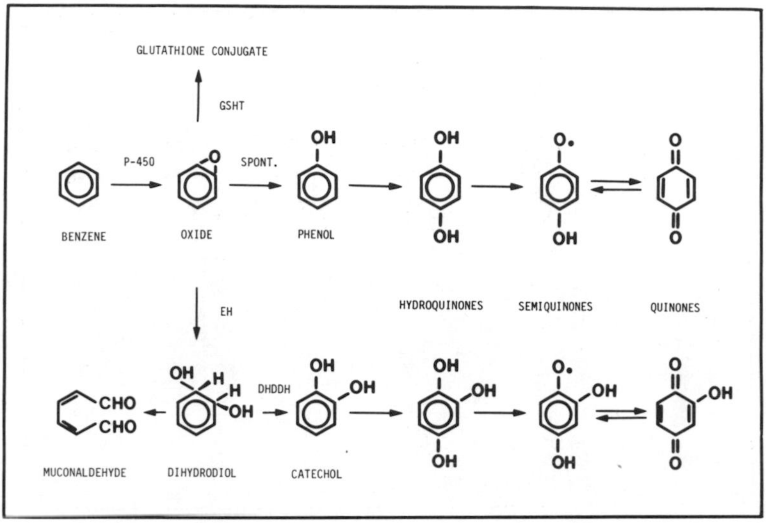

Figure 1. The metabolism of benzene. Abbrevations: P-450 = cytochrome P-450, EH = epoxide hydrolase, GSHT = glutathione S-transferase, DHDDH = dihydrodiol dehydrogenase.

The metabolism of styrene is shown in figure 2. The main reactive intermediate is believed to be the side chain epoxide formed by the action of cytochrome P-450 or some other oxidizing systems. In addition the role of the arene (ring-) epoxides has been denoted in certain instances. Also the importance of the aldehydes is somewhat obscure. For references see eg. Järvisalo 1978, WHO 1983.

Chloroform and carbon tetrachloride

The main toxic effects of chloroform are directed to the liver and kidney. Carbon tetrachloride is a model hepatotoxin.

The most apparent reactive intemediate of chloroform is phosgene but trichloromethyl radical and dichloromethyl carbene may be involved in extrahepatic metabolism. In the liver cytochrome P-450 is responsible for the metabolism see Hodgson et al. 1979. The reactive intermediate of carbon tetrachloride is trichloromethyl radical originating by the action of cytochrome P-450. This radical is able both to bind covalently and to start lipid peroxidation (see Jollow et al. 1977).

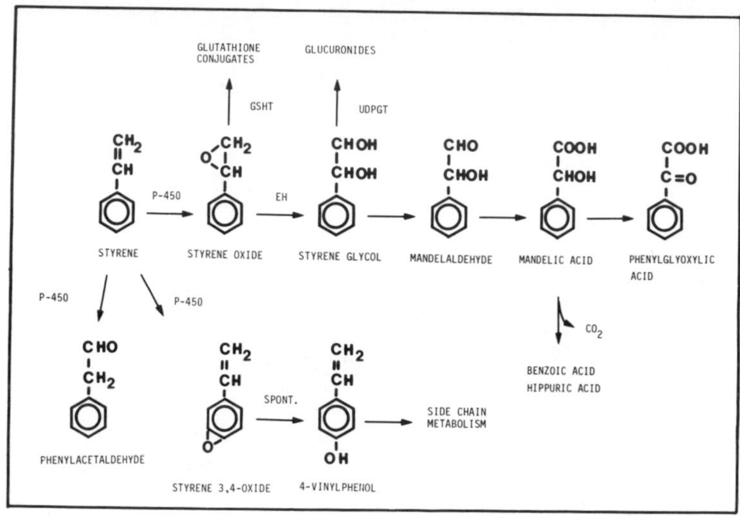

Figure 2. The metabolism of styrene. Abbrevations: see fig. 1, UDPGT = UDPglucuronosyltransferase.

Halogenated C-2-compounds

Haloalkanes are metabolized to aldehydes and acylhalides which represent the reactive forms of these compounds (See Hodgson et al. 1979). In alkenes the double bond is first epoxidated but on further spontaneous rearrangements the epoxides form aldehydes (see Snyder et al. 1982).

1,2-Dichloroethane is a hepato- and renotoxin. It is metabolized by cytochrome P-450 via chloroacetaldehyde, which is the reactive species. Also a glutathione conjugate has been proposed as a reactive intermediate (Guengerich et al. 1980) (See also IARC Monographs 20, 1979).

1,1,2-Trichloroethane and 1,1,2,2-tetrachloroethane are 10-20-fold more toxic than the ethylene congeners (tri- and tetrachloroethylene). 1,1,2-Trichloroethane is hepato- and renotoxic, 1,1,2,2-tetrachloroethane is a potent hepatotoxin. The metabolism of trichloroethane is assumed to proceed via chloroacetaldehyde. Minor metabolic routes of tetrachloro ethane go via tetrachloroethylene, the main route leading to dichloroacetic acid. The apparent reactive intermediate thus is dichloroacetylchloride. The metabolism of tri- and tetrachloroethylenes involves epoxidation. The reactive intermediates in vivo have not been identified with certainty (see IARC Monographs No 20, 1979).

Vinyl chloride causes acutely liver damage. It is a recognized human carcinogen causing angiosarcomas (IARC Monographs No 7, 1974). Vinyl chloride is metabolized by the microsomal monooxygenases to chloroethylene oxide, which rearranges to chloroacetaldehyde. The latter apparently is the main toxic species (see ,for instance, Snyder et al. 1982).

The significance

The understanding of the role of metabolism and the formation of reactive intermediates has elucidated the mechanisms of toxicity of many apparently harmless compounds. It has been of particular relevance for irreversible organ damage, mutagenicity, carcinogenicity and related effects. The underlying knowledge has been applied eg. in modelling alkylating and radical-forming sytostatics. The mechanistic information is also of great use for the development of safer medical compounds, pesticides and industrial chemicals. Furthermore, it provides means for detecting and estimating potentially serious toxicity of chemicals not yet studied.

REFERENCES

Gillette JR, Mitchell, JR, Brodie, BB (1974). Biochemical mechanisms of drug toxicity. Ann Rev Pharmacol 14, 271-288.

Guengerich FP, Crawford WM, Domoradzki JY, Macdonald TL, Watanabe PG (1980). In vitro activation of 1,2-dichloroethane by microsomal and cytosolic enzymes. Toxicol Appl Pharmacol 55:303-317.

Hodgson E, Bend JR, Philpot RM (1979). Reviews in biochemical toxicology. Elsevier.

Jakoby WB (1980). Enzymatic basis of detoxication I-II, Academic Press.

Järvisalo J (1978). Proceedings of the International Symposium on Styrene: Occupational and Toxicological aspects. Scand J Work Environ Health 4, Suppl. 2.

Jollow DJ, Kocsis JJ, Snyder J, Vainio H (1977). Biological reactive intermediates: formation, toxicity and inactivation. Plenum Press.

Lu AYH, Miwa GT (1980). Molecular properties and biological functions of microsomal epoxide hydrolase. Ann Rev Pharmacol Toxicol 20:513-531.

McBrien DCH, Slater TF (1982). Free radicals, lipid peroxidation and cancer, Academic Press.

Miller EC, Miller JA (1947). The presence and significance of bound aminoazo dyes in the livers of rats fed-dimethylaminoazobenzene. Cancer Res 7:468-480.
Miller EC, Miller JA (1982). Reactive metabolites as key intermediates in pharmacologic and toxicologic responses: Examples from chemical carcinogenesis. In: R Snyder, DV Parke, JJ Kocsis, DJ Jollow, CG Gibson, CM Witmer, Eds. Biological reactive intermediates II Chemical mechanisms and biological effects, Plenum Press, N.Y. pp. 1-21.
Pryor WA (1982). Free radicals in biology. Vol. V. Academic Press.
Snyder R, Parke DV, Kocsis JJ, Jollow DJ, Gibson CG, Witmer CM (1982). Biological reactive intermediates II. Chemical mechanisms and biological effects. Plenum Press.
WHO (1983). Environmental health criteria 26, Styrene.

Safety and Health Aspects of Organic Solvents, pages 97–105

METABOLIC INTERACTIONS BETWEEN ORGANIC SOLVENTS AND OTHER CHEMICALS

Martin Døssing, MD

Medical Department F, Gentofte University Hospital, DK-2900 Hellerup, Denmark

INTRODUCTION

It is the exception rather than the rule that workers are exposed to one chemical at a time. Many chemicals are often present in the working environment. Paints, laquers, adhesives etc. usually contain a mixture of many different organic solvents, and a fairly large proportion of the working population is at daily exposure to organic solvents. Ethanol is a widely used industrial solvent, but even if a person is exposed to an ethanol concentration of 1900 mg/m^3 a whole day exposure will only correspond to one glass of wine. Accordingly, inhalation exposure to ethanol is probably of minor importance compared to ingestion of ethanol containing drinks. Moreover, a considerable part of the working population is exposed to drugs such as antihypertensives (beta-blockers), H_2-antagonists (cimetidine) for peptic ulcer , antiepileptics (carbamazepine, phenytoin, and phenobarbitone) etc. Accordingly, workers are often exposed to a variety of chemical agents and interactions between these agents may lead to changes in toxicity of the solvents and failure or toxicity of drug therapy.

It is likely that the basic mechanism behind many of these interactions is of metabolic origin, and occurs in the polysubstrate monooxygenase enzyme system (cyt p-450) mainly confined to the liver. However, interactions may occur at any step from the absorption to the excretion of the chemicals including processes at the target organ.

Fig. 1

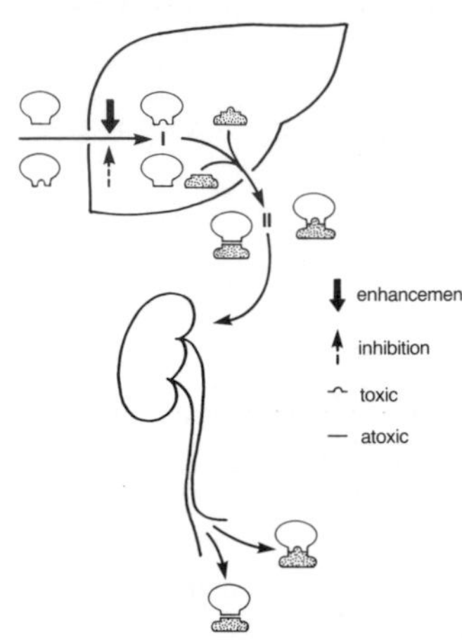

Fig. 1 is an oversimplification but meant to illustrate the effect of enhancement (induction) and inhibition of the metabolism of drugs and chemicals. The biotransformation involves two major steps, an oxidation reaction (phase I) and a conjugation reaction (phase II) with a small endogenous molecule such as glucuronic acid (indicated on the figure by the dotted symbols). If the parent compound is toxic (indicated by a special symbol on the figure) then an enhancement of the metabolism may decrease the systemic toxicity, and inhibition will increase the toxicity. The opposite is true if the parent compound is harmless, and a metabolic product or intermediate is toxic. The possible impact of changes in conjugation reactions has only been studied to a very limited extend in man.

The basic activity of the microsomal enzyme system of the liver is determined by genetic factors and environmental factors including smoking habits, dietary habits, physical activity, and diseases (Vesell 1982). While the combined effect of genetic and environmental factors are believed to be a major determining factor for the large interindividual difference in the metabolism of chemicals often amounting to 500%, changes in environmental factors are probably responsible for the rather large day-to-day variation (intraindividual variation) in enzyme activity amounting to 60-100% for some drugs (Nash et al. 1984).

In the working environment man is not only exposed to organic solvnets, chemicals, drugs, ethanol, but also to differences in diets, smoking habits etc. all of which may influence the activity of the microsomal enzyme system. It is further complicated by the fact that some environmental agents change different parts of the enzyme system to a different extent. Some agents may even initially serve as inhibitors of microsomal enzyme activity and later on act as inducers of the enzyme system. Also the time course of the changes may be of importance. Most inhibitors act by a competitive mechanism which occurs and subside as fast as the causative agents are eliminated from the body. But some inhibitors may destroy the microsomal enzyme system which has to be restored after withdrawal of the causative agent. Induction on the other hand often occurs after 7-10 days of exposure in man and subsides during 2-3 weeks free of exposure (Døssing et al. 1983). Exposure to more than one inducer or more than one inhibitor, or an inhibitor and an inducer at the same time may further complicate the outcome of metabolic interactions. Some inducers and some inhibitors act additively, some synergistically and some less than additively (Døssing 1982, Feely et al. 1983, Perucca et al. 1984).

The following review is an attempt to summarize the knowledge in this complicated research field of metabolic interactions in man. I will mainly deal with human studies because animal and human experiments often show conflicting results. For example short term exposure to ethanol induces the metabolism of solvents in rats (Sato et al. 1980), but inhibits solvent metabolism in man (Müller et al. 1975, Riihimäki et al. 1982, Wilson et al. 1983, Døssing et al. 1984). Accordingly, this review does not include information about interactions between occupational chemicals such as pesticides and solvents because no valid human data are yet available. It could be expected that the polychlorinated hydrocarbones such as the PCB's and some pesticides are inducers of solvent metabolism in man, but since phenobarbitone which is also a strong inducer of solvent metabolism in animals failed to change the metabolism of xylene in man (David et al. 1979), we have to await the results of human studies before conclusions can be made.

Solvent-solvent interactions

Few experimental investigations concerning possible interactions between solvents have been performed in man. At dose levels about the TLW's no interaction could be demonstrated after "a working day's" exposure to toluene and benzene (Sato et al. 1974), and styrene and acetone (Wigaeus and Nordquist 1984), while co-exposure to ethylbenzene and m-xylene resulted in a mutual metabolic inhibition as assessed by a delayed and decreased amount of urinary metabolic excretion (Engström et al. 1984). Also toluene and p-xylene had a modest inhibitory effect on the uptake and apparent clearance of each other (Wallén et al. 1985). In animals exposed to higher concentrations of solvents than those of the human studies a mutual inhibitory effect has been found between many solvent combinations i.e. benzene + toluene, styrene + toluene, m-xylene + toluene, trichloroethylene + toluene (summarized by Riihimäki 1984).

It has been known for decades that isopropanol, acetone, and ethanol may potentiate the hepatic and renal toxicity of the chlorinated solvents carbon tetrachloride, tetrachloroethylene, chloroform in rats (Cornish and Adefuin 1967, Traiger and Plaa 1971), and experiences from the working environment seem to indicate that the same holds true in man (Folland et al. 1976, Seeber et al. 1984). The mechanism behind this potentiation is generally ascribed to an inducing effect of the aliphatic alcohols on the formation of reactive metabolites from the chlorinated hydrocarbons which bind covalently to hepatic proteins (Ueng et al. 1983). This mechanism, however, hardly accounts for the synergism between trichloroethylene and carbon tetrachloride observed in young sniffers and confirmed in experiments with rats (Pessayre et al. 1982). Severe hepatotoxicity observed in workers without symptoms of CNS-intoxication during simultaneous exposure to many different solvents also indicate a synergistic metabolic interaction (Døssing and Ranek 1984).

The polyneuropathy associated with n-hexane and methyl n-butyl ketone is believed to be caused by a common metabolite, a γ-diketone, 2.5 hexanedione (Spencer et al 1980) However, the mechanism behind the aggravation of this neurotoxic reaction by methyl ethyl ketone is unexplained (Iwata et al. 1983).

Solvent-ethanol interactions

Exposure to carbon disulphide, dimethylformamide, and trichloroethylene may elicit ethanol intolerance with flushing, tachycardia, tachypnoe, sweat, etc., just like the disulfiram-ethanol reaction (Duric 1971, Chivers 1978, Stewart et al. 1974). This is probably due to inhibition of the aldehyde dehydrogenase leading to accumulation of acetaldehyde (Duric 1971, Lyle et al. 1979, Müller et al. 1975). Xylene and toluene have no effect on ethanol metabolism Riihimäki et al. 1982, Wallén et al. 1984).

Ethanol-solvent interactions

Table I shows that ethanol ingestion in socially accepted quantities has a profound inhibitory effect on the metabolism of trichloroethylene, styrene, xylene, and toluene (ref. in Table I).

Table I

Summary of metabolic ethanol-solvent interactions in man. Ethanol dose from 0.4 to 0.8 g/kg body weight giving a blood ethanol of about 6-21 mmol/l, solvent dose about or slightly above the TLW's.

SOLVENT	EFFECT ON SOLVENT BLOOD CONCENTRATION	EFFECT ON URINARY EXCRETION OF SOLVENT METABOLITES	EFFECT ON ETHANOL METANOLISM	REFERENCES
Trichloroethylene (TRI)	2-3 fold increase in TRI	0.6 fold decrease in trichloroacetic acid and trichloroethanol	1.1 fold increase in blood ethanol, 1.3 fold increase in blood acetaldehyde	Müller et al. 1975
Xylene	1.5-2 fold increase in xylene	0.5 fold decrease in methylhippuric acid, unchanged 2.4 xylenol	unchanged	Rihiimäki et al. 1982 Wilson et al. 1983
Styrene	15 fold increase in phenylethane 1.2 diol	0.6 fold decrease in mandelic acid	?	Wilson et al. 1983
Toluene	1.7 fold increase in toluene	0.5 times decrease in hippuric acid and O-cresol	unchanged	Waldron et al. 1983 Wallên et al. 1984 Døssing et al. 1984

The metabolic pathways of trichloroethylene and xylene are depressed differently, while the side chain and ring oxidation of toluene are inhibited equally. The inhibition of ethanol on solvent metabolism must be considered when interpreting results of biological monitoring based on urinary concentration of metabolites when the workers are exposed to trichloroethylene, styrene, xylene, and toluene.

Solvent-drug interactions

Exposure to carbon disulphide in concentrations as low as 10 ppm clearly depresses the N-demethylation of the model drug amidopyrine (Mack et al. 1974). The time course of this inhibitory effect indicates that it is competitive. Since carbon disulphide consistently inhibits the metabolism of microsomally metabolized drugs and chemicals in animals, it must be expected that carbon disulphide also acts as an inhibitor of solvent and drug metabolism in man.

In spray painters exposed to many solvents and other chemicals, the metabolism of the model drug antipyrine was depressed as assessed by a self administered one sample saliva method for antipyrine clearance determination (Døssing 1984). Exposure to mixtures of aliphatic and aromatic hydrocarbons (petrol, jet fuel) enhance antipyrine biotransformation (Harman et al. 1981, Døssing et al. (to be published)). In agreement with this the metabolism of warfarin was presumably stimulated in a house painter treated for deep vein thrombosis. This was reflected in differences in clotting factors during exposure to solvent based paints compared to exposure to water based paints (Lindberg et al. 1984). Long term exposure to styrene, toluene, and halothane about the TLW's did not change the clearance of antipyrine from saliva (Døssing 1984). This does, however, not exclude that the metabolism of other drugs may be altered by these solvents.

Drug-solvent interactions

While phenobarbitone is a strong inducer of solvent metabolism in experimental animals, it failed to influence the urinary excretion of both the conjugated and non-conjugated main metabolite of xylene in man (David et al. 1979). Also the well established inhibitors of microsomal drug metabolism cimetidine and propranolol did not change toluene metabolism in man in doses that clearly inhibit the metabolism of antipyrine (Døssing 1984).

DISCUSSION

Only the short term clinical outcome of solvent-ethanol-drug interactions is established. The long term consequences of these interactions are unknown. Is the solvent dose needed to cause chronic brain damage smaller in subjects with a moderate daily ethanol consumption than in people with negligible alcohol consumption? Or is less ethanol required for development of cirrhosis in workers with concomitant occupational exposure to solvents? It is also possible that these interactions may change the susceptibility to cancer.

REFERENCES

Chivers CP (1978). Disulfiram effect from inhalation of dimethylformamide. Lancet i:331.

Cornish HH, Adefuin J (1967). Potentiation of carbon tetrachloride toxicity by aliphatic alcohols. Arch Environ Health 14:447-449.

David A, Flek J, Frantik E, Gut I, Sedivec V (1979). Influence of phenobarbital on xylene metabolism in man and rats. Int Arch Occup Environ Health 44:117-125.

Duric D (1971). Some aspects of antabuse, carbon disulphide and ethyl alcohol metabolism. Arh Hig Rada 22:171-177.

Døssing M (1982). Changes in hapatic microsomal enzyme function in workers exposed to mixtures of chemical agents. Clin Pharmacol Ther 32:340-346.

Døssing M (1984). Noninvasive assessment of microsomal enzyme activity in occupational medicine. Int Arch Occup Environ Health 53:205-218.

Døssing M, Bælum J, Hansen SH, Lundqvist GR (1984). Effect of ethanol, cimetidine and propranolol on toluene metabolism in man. Int Arch Occup Environ Health 54:309-315.

Døssing M, Loft S, Schroeder E. Jet fuel and liver function (to be published).

Døssing M, Pilsgaard H, Rasmussen B, Poulsen HE (1983). Time course of phenobarbital and cimetidine mediated changes in hepatic drug metabolism. Eur J Clin Pharmacol 25:215-222.

Døssing M, Ranek L (1984). Isolated liver damage in chemical workers. Br J Industr Med 41:142-144.

Engström K, Riihimäki V, Laine A (1984). Urinary disposition of ethylbenzene and m-xylene in man following separate and combined exposure. Int Arch Occup Environ Health 54: 355-363.

Feely J, Pereira L, Guy E, Hockings N (1984). Factors affecting the response to inhibition of drug metabolism by cimetidine - dose response and sensitivity of elderly and induced subjects. Br J Clin Pharmacol 17:77-81.

Folland DS, Schaffner W, Ginn HE, Crofford OB, McMurray DR (1976). Carbon tetrachloride toxicity potentiated by isopropylalcohol. JAMA 236:1853-56.

Harman AW, Frewin DB, Priestly BG (1981). Induction of microsomal drug metabolism in man and the rat by exposure to petroleum. Br J Industr Med 38:91-97.

Iwata M, Takeuchi Y, Hisanaga N, Ono Y (1983). Changes of n-hexane metabolites in urine of rats exposed to various concentrations of n-hexane and to its mixture with toluene or MEK. Int Arch Occup Environ Health 53:1-8.

Lindberg E, Lundberg S, Hogstedt C, Hensjö LO (1984). Kan yrkesmässig lösningsmedelsexponering minska effekten av antikoagulantia? Läkartidningen 81:110.

Lyle WH, Spence TWH, McKinneley WM, Duckers K (1979). Dimethylformamide and alcohol intolerance. Bt J Industr Med 36 63-66.

Mack T, Freundt KJ, Henschler D (1974). Inhibition of oxidative n-demethylation in man by low doses of inhaled carbon disulphide. Biochem Pharmacol 23:607-614.

Müller G, Spassowski M, Henschler D (1975). Metabolism of trichloroethylene in man III. Interaction of trichloroethylene and ethanol. Arch Toxicol 33:173-189.

Nash RM, Stein L, Penno MB, Passananti GT, Vesell ES (1984). Sources of interindividual variations in acetaminophen and antipyrine metabolism. Clin Pharmacol Ther 36:417-430.

Peruca E, Hedges A, Makki KA, Ruprah M, Wilson JF, Richens A (1984). A comparative study of the relative enzyme inducing properties of anticonvulsant drugs in epileptic patients. Br J Clin Pharmacol 18:401-410.

Pessayre D, Cobert B, Descatoire V, Degott C, Babany G, Funk-Bretano C, Delaforge M, Larrey D (1982). Hepatotoxicity of trichloroethylene-carbon tetrachloride mixtures in rats. Gastroenterology 83:761-772.

Seeber A, Dietz E, Gutewort T, Zeller H-J (1984). Tetrachlotoethylene exposure and ethanol consumption - behavioral, neurophysiological and biochemical effects. In International Conference on Organic Solvent Toxicity. Stockholm, Sweden p 383.

Spencer PC, Schaumburg HH, Sabri MI, Veronesi B (1980). The enlarging view of hexacarbon neurotixicity. CRC Crit Rev Toxicol 7:279-357.

Riihimäki V (1984). Interactions between industrial solvents and between solvents and ethanol. In Aitio A, Riihimäki V, Vainio H (eds.): "Biological Monitoring and Surveilance of Workers Exposed to Chemicals", Washington, New York, London: Hemisphere Publishing Corporation pp 231-243.

Riihimäki V, Savolainen K, Pfäffli P, Pekari K, Sippel HW, Laine A (1982). Metabolic interaction between m-xylene and ethanol. Arch Toxicol 49:253-263.

Sato A, Nakajima T, Fujiwara Y, Hirosawa K (1974). Pharmacokinetics of benzene and toluene. Int Arch Arbeitsmed 33:1 169-182.

Sato A, Nakajima T, Koyama Y (1980). Effects of chronic ethanol consumption on hepatic metabolism of aromatic and chlorinetad hydrocarbons in rats. Br J Industr Med 37:382-386.

Stewart RD, Hake Cl, Peterson JE (1974). Degreasers flush. Arch Environ Health 29:1-5.

Traiger GJ, Plaa GL (1971). Differences in the potentiation of carbon tetrachloride in rats by ethanol and isopropanol pretreatment. Toxicol Appl Pharmacol 20:105-112.

Ueng T-H, Moore L, Elves RG, Alvares AP (1983). Isopropanol enhancement of cytochrome p-450-dependent monooxygenase activities and its effects on carbon tetrachloride intoxication. Toxicol Appl Pharmacol 71:204-214.

Vesell ES (1979). The antipyrine test in clinical pharmacology: conceptions and misconceptions. Clin Pharmacol Ther 26:275-286.

Wallén M, Holm S, Nordqvist MB (1985). Coexposure to toluene and p-xylene in man: uptake and elimination. Br J Industr Med 42:111-116.

Wallen M, Näslund PH, Nordqvist MB (1984). The effect of ethanol on the kinetics of toluene in man. Toxicol Appl Pharmacol 76:414-419.

Wigaeus E, Löf A, Nordqvist MB (1984). Uptake, distribution, metabolism, and elimination of styrene in man. A comparison between single exposure and co-exposure with acetone. Br J Industr Med 41:539-546.

Wilson HK, Robertson SM, Waldron HA, Compertz D (1983). Effect of alcohol on the kinetics of mandelic acid excretion in volunteers exposed to styrene vapour. Br J Industr Med 40:75-80.

Safety and Health Aspects of Organic Solvents, pages 107–114

HOST SUSCEPTIBILITY IN ORGANIC SOLVENT TOXICITY

Olavi Pelkonen

Department of Pharmacology, University of Oulu, SF-90220 Oulu 22, Finland

FACTORS AFFECTING SUSCEPTIBILITY TO TOXIC REACTIONS

Toxic reaction elicited by a chemical is dependent on the inherent toxicological profile of the substance, on the exposure situation, and on the host. It is not an uncommon finding that any measured outcome of an exposure to a foreign chemical, be it LD50, tissue injury or cancer, is strongly dependent on the species, strain and individual under study. Exposed to similar amounts of cigarette smoke over years, one individual contracts lung cancer, another does not.

Interindividual differences in the action of toxic substances are due to three types of factors (defined in a loose way):

1) Genetic host factors

These are the "commands" in the genome of an individual. Sometimes it is difficult to make a distinction between genetic and non-genetic host factors. For example, although variations caused by age are certainly based on genetic "programs", age is nevertheless treated in most instances like a non-genetic host feature of an organism, undoubtedly because we know very little about those genetic "programs".

2) Non-genetic (acquired) host factors

These are non-genetic attributes of organisms, variation in which is associated with variation in the risk of developing toxicity. A comprehensive list of different host factors is difficult to compile, because of the somewhat ambiguous meaning of the term, but a good starting point can be found in the article of Vesell (1983).

3) Environmental factors
These are any exogenous influences, be they physical, chemical, microbial or viral, on an organism. Usually these influences are temporary, but it must be noted, that sometimes they may lead to long-lasting or even permanent changes in the organism, in other words, to "host factors".

In real life, it is usually difficult to elucidate, which factors are the most significant ones in affecting the toxic response under study. The genetic background represents only the potential of an individual and sets the limits to the response, but the full potential to respond is revealed only at real life situations, as a result of acquired host factors and environmental exposures.

This article is intended to be a short introduction to the analysis of the role of host factors in toxic reactions, particularly in connection with organic solvent exposure. However, it is worth of stating, that very little "hard" data exist concerning the role of host in organic solvent toxicity.

TOXICOKINETICS AND TOXICODYNAMICS

The toxicological bases of individual variability are at either toxicokinetic or toxicodynamic levels. The concentration of an active substance at the site of action is governed by toxicokinetic processes: absorption, distribution, metabolism and excretion. Although there are often rather wide variations in absorption, distribution and excretion, the most significant variation occurs in metabolism. Almost all foreign chemicals are metabolized in the body by the catalysis of xenobiotic-metabolizing enzymes and the activity of these enzymes displays enormous interindividual variability (table 1). This variation leads to comparable differences in the concentration of a toxic substance at the site of action.

EXAMPLES OF HOST SUSCEPTIBILITY IN TOXIC REACTIONS

Genetic host factors

The most significant toxicogenetic conditions have been revealed with respect to xenobiotic-metabolizing enzymes probably because these enzymes can be rather easily measured both <u>in vitro</u> and <u>in vivo</u> and in certain cases the consequence of the defect is so conspicuous, e.g. apnea after the administration of succinylcholine in patients with atypical cholinesterase unable to hydrolyze choline esters. Other well-known examples include polymorphisms in debrisoquine hydroxylation, benzo(a)pyrene oxidation (aryl hydrocarbon hydroxylase) and acetylation of aromatic amines. It is probable that many more genetic conditions affecting xenobiotic metabolism will be elucidated in near future, because nowadays many sophisticated analytical techniques, for example, the use of monoclonal antibodies against xenobiotic-metabolizing enzymes for "phenotyping" individual patterns of these enzymes, are becoming available.

TABLE 1. Some Examples of the Interindividual Variability in the Activity of Xenobiotic-metabolizing Enzymes

Enzyme/metabolic reaction	Variation	Remarks
Debrisoquine 4-hydroxylation		
Urinary ratio for 4OHD/D	400	Polymorphic
Activity in vitro	Poor metabolizers: no high affinity component	
Antipyrine elimination		
clearance	40	Unimodal
metabolite formation	10-30	Trimodal
half-life	25	Unimodal
Coumarin 7-hydroxylase		
activity in vitro	150	
Aryl hydrocarbon hydroxylase		
activity in liver biopsies	30-80	Smoking: small effect
in placentas	350-1000	Strongly dependent on maternal smoking
in lung biopsies	20- >50	Smoking: no effect
induction in lymphocytes	5-10	Trimodal (?)

Variation is expressed as a ratio between the highest and the lowest value detected and it is strongly dependent on the characteristics of a population under study. Thus actual figures for variation differ in different studies, but above figures are mainly representative of the author's experiences.

Much less is known about inherited interindividual variability at the toxicodynamic level. It is firmly established that variations in receptor number and sensitivity in various diseases and drug exposures exist. It is also probable that there are both monogenic and polygenic variations in the structure and amount of receptors and other components in the receptor-effector systems. Few monogenically inherited pharmacodynamic conditions are known and in most cases the molecular basis has not yet been elucidated.

Non-genetic host factors

It is clear that the genetic background is just a possibility; its expression is dependent on the intact recognition and effector systems. These systems can be transiently or permanently affected by changes in the body's homeostasis: for example, hormone balance or diseases. An example of the changes which could become permanent is liver disease and its effects on drug metabolism: early changes, such as fatty infiltration and mild inflammation, are basically reversible, and also depression of xenobiotic metabolism, if it occurs at all, is mild and reversible. Later and more severe manifestations, such as hepatitis and cirrhosis cause partially irreversible changes in the structure and function of the liver. Although the cirrhotic liver still can respond to exogenous chemicals and express at least some activity at the decreased level, it can never display again the original potential of the genetic background.

Environment

However, the final outcome is dependent on the interplay between genetic, host and environmental factors. Numerous studies in clinical pharmacology and occupational toxicology have demonstrated that people exposed to drugs, occupational chemicals, pesticides or any other inducing substances metabolize model compounds faster than non-exposed people and still there is a wide variation in the extent of induction. At the current level of knowledge, it is difficult or even impossible to elucidate whether the extent of induction is determined by the inherent ability of an individual to be induced (determined basically by genetic and host factors) or by the dose of an environmental chemical to which an individual has been exposed.

HOST SUSCEPTIBILITY IN ORGANIC SOLVENT TOXICITY

Toxic effects of solvent exposure

Very simplistically, the toxic effects of organic solvents can be divided into acute and delayed effects (Table 2). Probably the mechanisms behind these two kinds of toxic outcomes are different. Acute effects, which are primarily manifestations of CNS toxicity, for example narcosis, are probably very similar to those caused by anesthetic agents and are said to be due to non-specific membrane interactions of lipid-soluble substances or to some kind of receptor interactions. Delayed effects, damages of liver and other parenchymal tissues and delayed neurotoxicity, are probably caused by the production of reactive intermediates and the covalent binding of these intermediates with critical cellular macromolecules.

TABLE 2. Systemic Toxic Effects of Organic Solvents

	Probable mechanism	Role of host
Acute effects		
CNS depression	membrane-mediated	probably no
Cardiac arrest	receptor-mediated	not known
Delayed effects		
Encephalopathy	not known	not known
Peripheral neurotoxicity	metabolism-mediated	not known
Parenchymal tissue damage	metabolism-mediated	probably yes
Teratogenicity	not known	not know
Mutagenicity	metabolism-mediated	probably yes
Carcinogenicity	metabolism-mediated	probably yes

Solvent toxicity and host susceptibility

Ethanol may serve as an example for other organic solvents, because - for obvious reasons - it is the most thoroughly studied substance among them. Ethanol is metabolized by alcohol dehydrogenase and aldehyde dehydrogenase, both of which display remarkable polymorphism. This leads to a large variety of individually and racially different enzyme phenotypes. A particular phenotype may make its carrier vulnerable to alcohol-induced toxicities. For example, some Orientals get severe acute symptoms after ingestion of small amounts of ethanol. This sensitivity is associated with the specific phenotype of aldehyde dehydrogenase.

Ethanol is also metabolized by the so called "microsomal ethanol-oxidizing system" (MEOS), which is inducible by ethanol itself and a number of other chemicals. MEOS metabolizes also other solvents, e.g. higher alcohols and acetone. To what extent the activity and inducibility of this enzyme system is regulated by host factors and what is its role in ethanol-induced toxicity, is not known.

Otherwise, very little is known about the significance of host factors in solvent toxicity. Methyl chloride is known to be excreted as S-methylcysteine and there seem to exist two groups of people with respect to that capacity: "poor" and "extensive" converters (van Doorn et al., 1980). However, with most organic solvents there exist very little relevant information and what follows is mostly speculation. With respect to acute toxicity, mediated by non-specific or receptor interactions, very little is known about underlying causes of interindividual variability. It is even generally argued, that interindividual variability in acute organic solvent toxicity is relatively small and that there are no really significant "idiosyncrasies" in these effects. The dose seems to be the most important single determinant of toxicity.

With respect to delayed toxic effects, experience in general toxicology points to the possible importance of host factors. As said above, especially in xenobiotic metabolism a very large interindividual variation has been observed and the activity of metabolizing enzymes may be of critical significance in both metabolic toxification and inactivation of organic solvents. Animal experiments have amply demonstrated the importance of metabolism especially in the delayed toxic effects of halogenated organic solvents and it is probable that this is true also in man. However, it is notoriously difficult to differentiate between liver injury caused by low, but sufficient, doses of organic solvent and liver injury caused in a susceptible individual by doses which are safe in "normal" individuals. The same fact holds true with other delayed effects.

SUMMARY

The importance of host factors, both genetic and non-genetic, is evident in the actions of foreign chemicals. Host factors may act at both toxicodynamic and toxicokinetic levels. Perhaps the best understood phase of influence is the metabolism of foreign substances, which displays enormous interindividual variation. Genetic, non-genetic and environmental factors are known to be behind this variation, although it is not always in real life easy to tell the relative importance of these different factors.

With respect to organic solvent toxicity, there are very little data about the significance of host factors. It is possible, however, that host susceptibility is of critical importance in delayed toxic manifestations of halogenated organic solvents. This view is based mainly on animal experiments.

ACKNOWLEDGEMENTS

The author wish to thank Dr. Kirsi Vähäkangas and Dr. Vesa Riihimäki for useful comments.

RELEVANT LITERATURE

General aspects

Sato R, Kato R (eds) (1982). "Microsomes, Drug Oxidations, and Drug Toxicity". Tokyo: Japan Sci Soc Press.

Boobis AR, Caldwell J, De Matteis F, Elcombe CR (eds) (1985). "Microsomes and Drug Oxidations". London: Taylor & Francis.

Atlas SA, Nebert DW (1977). Pharmacogenetics and Human Disease. In Parke DV, Smith RL (eds): Drug Metabolism: from Microbe to Man," London: Taylor & Francis, pp. 393-430.

Vesell ES (1983). On the significance of host factors that affect drug disposition. Clin Pharmacol Ther 31: 1-7.

Harris CC, Mulvihill JJ, Thorgeirsson SS, Minna JD (1980). Individual differences in cancer susceptibility. Ann Int Med 92: 809-825.

Bartsch H, Armstrong B (eds) (1982). "Host Factors in Human Carcinogenesis". Lyon: IARC Sci Publ No. 39.

Turusov V, Montesano R (eds) (1983). "Modulators of Experimental Carcinogenesis". Lyon: IARC Sci Publ No. 51.

Organic solvent toxicity

Aitio A, Riihimäki V, Vainio H (eds) (1984). "Biological Monitoring and Surveillance of Workers Exposed to Chemicals". Washington: Hemisphere.

van Doorn R, Borm PJA, Leijdekkers CM, Henderson PT, Reuvers J, van Bergen TJ (1980). Detection and identification of S-methylcysteine in urine of workers exposed to methyl chloride. Int Arch Occupat Environ Health 46: 99-109.

Gompertz D (1981). Solvents. The relationship between biological monitoring strategies and metabolic handling. A review. Ann Occupat Hyg 23: 405-410.

Toftgård R, Gustafsson J-Å (1980). Biotransformation of organic solvents. A review. Scand J Work Environ Health 6: 1-18.

von Wartburg J-P, Buhler R (1984). Biology of disease. Alcoholism and aldehydism: New biomedical concepts. Lab Invest 50: 5-15.

Safety and Health Aspects of Organic Solvents, pages 115–119

EFFECTS OF ORGANIC SOLVENTS ON LIVER CELL MORPHOLOGY

Juha Nickels

Department of Occupational Medicine,
Institute of Occupational Health,
Helsinki, Finland

The target organelle of many xenobiotics in the hepatocyte is the smooth endoplasmic reticulum (SER). This membrane complex harbors a complicated enzyme system, containing the elements for mixed function oxidase (MFO) reactions, e.g. NADPH-cytochrome C reductase, Cytochrome P-450 and UDPGA-transferase (Dallner and Ericsson, 1976).

Some industrially used solvents (benzene, xylene, toluene) are MFO-enzyme system inducers but do not cause hepatic necrosis as do some other hepatotoxins, e.g. vinyl chloride, carbon tetrachloride, carbon disulfide, trichlorethylene, tetrachloroethylene and trinitrotoluene (Reynolds, 1977). Some solvents, on the other hand, may potentiate hepatotoxicity of other chemicals or interact with other solvents, such as ethanol. When properly used, these solvents are considered safe, and no adverse effects on liver enzyme tests have been demonstrated in well controlled studies (Kurppa and Husman, 1982).

Experiments on animals, however, indicate that some solvents may have injurious effects on liver cells not discovered by ordinary qualitative light microscopical methods. Combined light microscopical morphometry and electron microscopy are more sensitive tools when the effects of solvents or other xenobiotics are studied. Vinyl toluene inhalation has no or only slight effect on the biochemical activities of hepatic drug biotransformation enzymes (Heinonen, Nickels and Vainio, 1982) but distinct histological changes are visible by these methods (Table 1). The cell volume (area) changes

TABLE 1. Effect of vinyltoluene inhalation (300 ppm) on the mean size of hepatocytes. Number of cells in parentheses.

Duration (weeks)	Cell size (μm^2) (mean $\pm$ S.D.)	
	Control	300 ppm
1	459 $\pm$ 93 (66)	396 $\pm$ 108 (45)**
4	475 $\pm$ 117 (41)	505 $\pm$ 147 (56) NS
8	458 $\pm$ 118 (65)	323 $\pm$ 83 (61)***
12	463 $\pm$ 116 (27)	405 $\pm$ 116 (44)*
15	458 $\pm$ 141 (31)	384 $\pm$ 104 (40)*

NS=not significant, *$P<0.05$, **$P<0.01$, ***$P<0.001$

significantly both in acute and subchronic experiments. Quantitative morphometry appears to be a sensitive method in these experiments showing both oscillatory and adaptive mechanisms in liver cells after solvent treatment (Fig.1).

Polychlorinated biphenyl (PCB) pretreatment compensates the acute effects of vinyltoluene on liver cell volume (area) by proliferating the SER (Table 2). The volume changes of hepatocytes in these experiments depend on the great proliferating capacity of SER. This compensates the great MFO-enzyme demand exerted on the hepatocytes by the solvents. This capacity seems to be reduced in subchronic exposure, a sign of possible exhaustion.

Application of this stereological method to human liver biopsies is difficult because of the great variation in normal human environment, habits and constitution. Morphometry may, on the other hand, be the only present method by which an effect

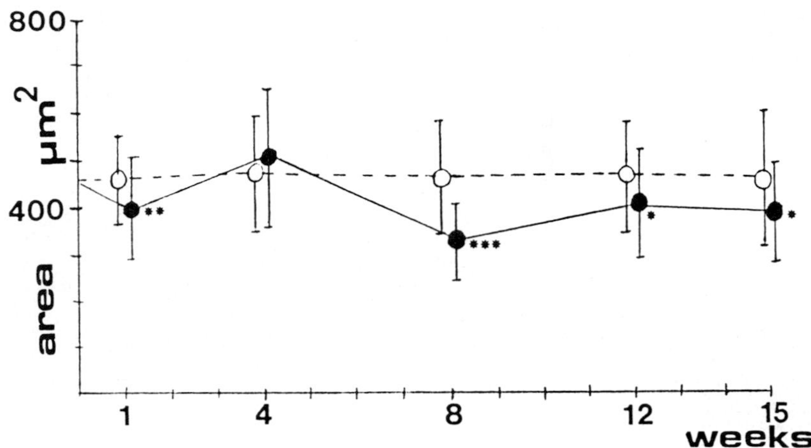

Figure 1. Area of rat hepatocytes after 300 ppm vinyltoluene (•) exposure, (o = control). *P< 0.05, **P< 0.01, ***P< 0.001

TABLE 2. Effects of vinyltoluene inhalation (300 ppm, 1 week) and/or a single injection of PCBs (500 mg/kg 5 days before) on the mean size of hepatocytes. Number of cells in parentheses.

Exposure	Cell size (um^2) (mean ± S.D.)
Control	459 ± 93 (66)
Vinyltoluene	396 ± 108 (45)**
PCBs	538 ± 148 (76)***
Vinyltoluene + PCBs	469 ± 110 (41) NS

NS=not significant, **P<0.01, ***P<0.001

on human liver cells can be seen after daily industrial solvent exposure.

Hepatic drug oxidases are under genetic control (Kalow and Inaba, 1976). Therefore one can expect to find individuals, who react abnormally after exposure to occupational solvents. Possible

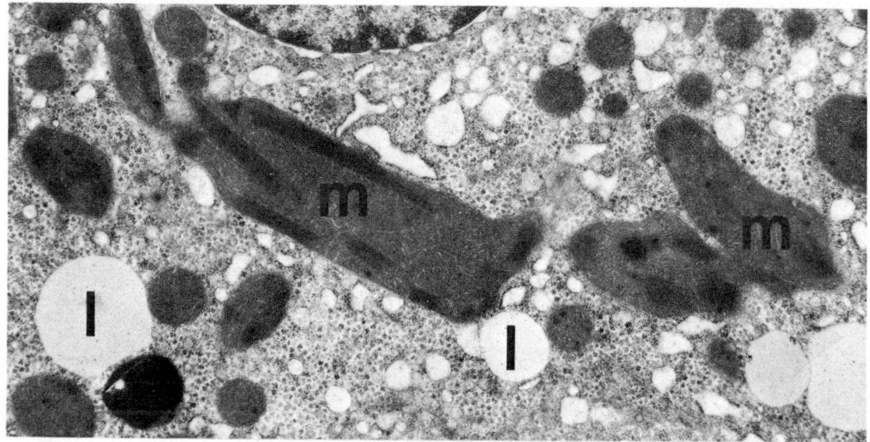

Figure 2. Electron micrograph of a liver cell from a 55 year old laundry-keeper exposed to perchlorethylene for 3 years. Giant mitochondria (m) and lipid (l) vacuoles point to ethanol injury. The patient consumed 1-2 bottles of wine weekly. Liver enzymes (U/l): ASAT 51-114, ALAT 90-128, gamma-GT 61-186. Magnification X 11 700.

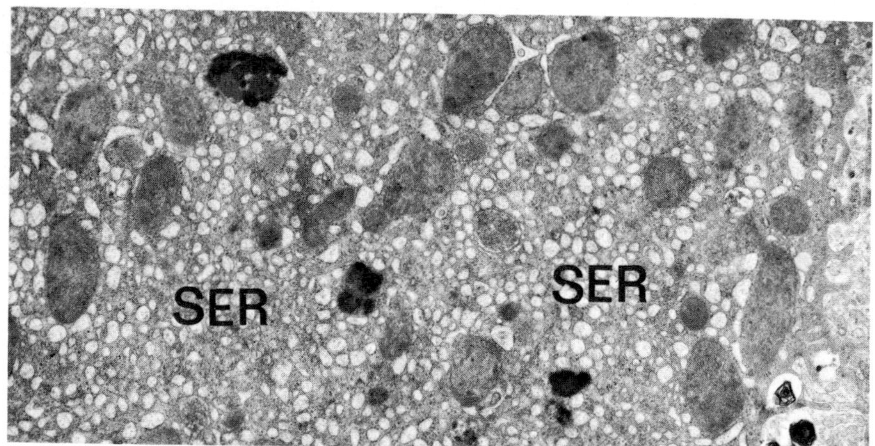

Figure 3. Electron micrograph of a liver cell from a 32 year old painter exposed to solvents 1-2 hours/day for 5 years. Mitochondria are normal but SER proliferates as a sign of solvent effect. Liver enzymes (U/l): ASAT 31-60, ALAT 67-149, gamma-GT 115-211. Magnification X 8 800.

liver damage is, however, difficult to prove in these cases. In daily practice and in the absence of more sensitive liver function tests, detailed occupational history combined with thorough clinical evaluation and liver biopsy remains the mainstay in the evaluation of environmental injuries. Using light and electron microscopy the differential diagnosis between alcoholic liver disease, e.g. steatosis with giant mitochondria (Fig. 2) and injury following exposure to industrial solvents, e.g. steatosis and proliferative SER (Fig.3) may be possible. Histological changes resembling viral hepatitis (either chronic persistent hepatitis or chronic active hepatitis) and cholestasis point to other causes than industrial solvents.

REFERENCES

Dallner G, Ericsson L (1976). Molecular structure and biological implication of the liver endoplasmic reticulum. Progress in Liver Dis 5:35.

Heinonen T, Nickels J, Vainio H (1982). Subacute toxicity of vinyltoluene vapour: Effects on the hepatic and renal drug biotransformation and the urinary excretion of thioether. Acta Pharmacol Toxicol 51:69.

Kalow W, Inaba T (1976). Genetic factors in hepatic drug oxidations. Progress in Liver Dis 5:246.

Kurppa K, Husman K (1982). Car painters' exposure to a mixture of organic solvents. Serum activities of liver enzymes. Scand j work environ health 8:137.

Reynolds E (1977). Environmental aspects of injury and disease: liver and bile ducts. Environ health persp 20:1.

OCCUPATIONAL TOXICOLOGY

Safety and Health Aspects of Organic Solvents, pages 123–131

ACUTE SOLVENT INTOXICATION

Arto Laine and Vesa Riihimäki

Department of Industrial Hygiene and Toxiology
Institute of Occupational Health
SF-00290 HELSINKI, Finland

INTRODUCTION

Acute solvent intoxications were first documented in the medical literature for more than a hundred years ago. Although hygienic conditions at the workplace have vastly improved since that time and the mean exposure levels to solvents are still declining, acute solvent intoxications can and do occur still today. In Finland the average of one case of a fatal solvent intoxication has been encountered every year and several other hazardous incidents have occured in which the more serious consequences were avoided thanks to effective rescue operations or good luck.

The hazardous situations arise in typical ways: unprotected workers enter tanks or other confined spaces for cleaning, painting or repair purposes without prior assessment of the enviromental contaminants and without effective ventilation or protection during the work which releases solvent vapors. In accidental conditions the hazardous exposures may be caused by an overflow of tanks or an unexpected discharge of vapors from broken valves or pipes.

Acute solvent intoxications are mainly gassings in which the central nervous system (CNS) becomes depressed so insidiously or rapidly that the victim finds himself in a helpless condition and then loses conciousness.

EPIDEMIOLOGY OF ACUTE SOLVENT INTOXICATIONS

Benzene was isolated from coal tar naphta in 1825 starting the industrial history of solvent exposure (Gerarde, 1960). Probably the first case of an acute and clinically defined benzene poisoning was described in 1862 (Winslow, 1927). Since then acute benzene poisoning was considered one of the major problems of industrial hygiene over many decades. For example, in a summary of the National Safety Council study of acute benzene poisonings, Winslow (1927) reviewed the literature published in America and Europe before 1927 and obtained a record of about 200 cases of acute benzene poisoning of which some 70 were fatal.

More recently, other aromatic and halogenated aliphatic hydrocarbons have been reported to cause numerous cases of gassings, sometimes with fatal outcomes. Automotive fuels, mineral spirits and other petroleum solvents have caused several poisonings with primarily narcotic symptoms and signs. Alcohols, acetates and ketone solvents may likewise cause nervous system depression and irritation by inhalational route but serious acute intoxications from their industrial use are apparently rare (Browning, 1965; Brugnone et al; 1983, Longley et al., 1967; Morley et al., 1970; Wang and Irons, 1961).

McCarthy and Jones (1983) presented a statistical review of 384 cases of industrial gassings by trichloroethylene, perchloroethylene and 1,1,1 -trichloroethylene reported to HM factory inspectorate during the years 1961 - 1980 in the UK. 75 % of the cases involved trichloroethylene, 11 % perchloroethylene, and 14 % 1,1,1-trichloroethane. 4.5 % of the intoxications ended fatally and slightly less than a half of the incidents were severe enough to cause unconsciousness. Trichloroethylene caused 12 deaths, perchloroethylene 3, and 1,1,1 -trichloroethane the remaining 2. Solvent abuse was mentioned as a possible causative factor in nine cases altogether. A similar more recent review dealt with industrial gassings caused primarily by aromatic solvents in 1961 through 1980 in the U.K. (Bakinson and Jones, 1985). Of the reported cases 28 % were attributed to methylene chloride, 31 % to toluene, 32 % to xylene and 8.5 % to styrene exposures. The clinical manifestations again mainly concerned the central nervous system with 48 people becoming unconscious, and four persons died. Solvent addiction was found in one case.

Headache, lightheadedness, drowsiness, fatigue and dizziness are typical symptoms of central nervous system depression. Tremors and convulsions are more rare neuroirritant effects but clinically well known for benzene. Sensations of tightness of the chest and difficulties of breathing are also common. They may arise from a straightforward irritation of the upper respiratory tract. Severe inhalation poisonings by solvents with unconsciousness may be accompanied with lung oedema (Morley et al., 1970) and, if the liquid is aspirated, a serious chemical pneumonitis may ensue (Lee and Seymour, 1979). Nausea and functional disturbances of the gastrointestinal tract have also been described.

Inhalation of the usual aromatic hydrocarbon solvents, toluene, xylenes and styrene, has been reported to cause in individual cases narcotic effects ranging from a mild fatigue, dizziness and lightheadedness to serious intoxications with unconsciousness and death. Occasionally slight liver and kidney injuries have been observed in the more serious intoxications with prolonged unconsciousness (Browning, 1965; Gerarde, 1960; Morley et al., 1970).

All the commonly used halogenated hydrocarbons possess narcotic properties, and especially chloroform and trichloroethylene have been widely used as anaesthetic agents and analgesics. Some of the halogenated hydrocarbons are remarkably toxic to the liver and the kidney. Carbon tetrachloride is particularly notorious as a cause of acute liver and kidney injury. Trichloroethylene and tetrachloroethylene may cause acute liver and/or kidney injury in very high concentrations (Chenoweth and Hake, 1969; Torkelson and Rowe, 1982). Trigeminal nerve appears to be particularly suspectible to trichloroethylene, and in dental practice trichloroethylene has been used for pain relief. In some cases of high accidental industrial exposure trichloroethylene has caused anaesthesia of the facial skin with a slow recovery (Feldman et al., 1970). Methylene chloride is metabolized to carbon monoxide; thus it can impair cardiac oxygenation among workers who suffer ischemic heart disease (Stewart and Hake, 1976). The risk of cardiac arrhythmias has been attributed principally to high exposures involving fluorinated hydrocarbons, 1,1,1 -trichloroethane, trichloroethylene and toluene. Illustrative case reports mainly deal with the so-called sudden sniffing death-syndrome in which intentional inhalation of high concentrations of organic solvents followed by an emotional or physical stress led to a sudden death. The mechanism of death in a typical case is thought to be a solvent induced cardiac sensitization to catecholamines (Bass, 1970; Droz et al., 1982).

In exposures to solvent mixtures potentiation of effects might occur. Human case histories point out the hazards of the combined exposure to carbon tetrachloride and ingestion of alcohol. A report of a cluster of cases of acute industrial intoxication, some with renal involvement, in an isopropanol bottling facility implicated the combined exposure to vapors of isopropanol and carbon tetrachloride (Folland et al., 1976).

CORRELATION OF EXPOSURE AND EFFECT

Toluene, a model substance for aromatic solvents, caused over a few hours at the level of 400 ppm mild sensations of nausea and fatigue, and a slight incoordination was recorded (von Oettingen et al., 1942). At the level of 600 to 800 ppm a transient exhilaration was observed followed by a progressive fatigue, dizziness and weakness, and a moderate ataxia. At an exposure to 800 ppm over eight hours fatigue was accompanied by incoordination and signs of a staggering gait. Experience from incidents of toluene abuse and accidental overexposure indicates that concentrations exceeding 2000 to 3250 ppm may be fatal within 30 minutes (Nomiyama and Nomiyama, 1978). At an ambient toluene concentration of 5000 ppm or more acute narcotic symptoms may develop very rapidly, and the exposure may become life-threatening in minutes (Longley et al., 1967). The minimum anesthetic concentration for xylene was estimated to be around 5000 ppm on the basis of an accidental exposure (Morley et al., 1970).

The aromatics have a recognizable odor which is more or less unpleasant. At concentrations ranging from 400 to 800 ppm, toluene and xylene caused rather mild irritation of the eyes and the respiratory tract and even relatively high concentrations would not be irritating enough to preclude man from working (Carpenter et al., 1975; Gerarde, 1960; Longley et al., 1967). The odor of styrene is readily detectable between 60 - 100 ppm without causing discomfort and at 500 to 600 ppm strong irritation of the eyes and throat occurs. When the concentration of styrene was 800 ppm, an exposure over some hours caused drowsiness and unsteadiness. Irritation was immediately felt at that concentration and it became intolerable in some minutes to unacclimatized subjects (Wolf et al., 1956).

1,1,1 -trichloroethane has a typical sweetish odor which is usually noticeable at about 100 ppm. The odor was not regarded as unpleasant at 500 ppm and even 1000 ppm was tolerated without excessive irritation (Torkelson et al., 1958). Feelings of lightheadedness and a slight loss of coordination and equilibrium were described in volunteers who were exposed to 1,1,1 -trichloroethane over 1 hour at about 900 ppm. Depressant effects became more obvious, with a positive Romberg´s test, at 1900 ppm for 5 minutes, and the odor was strong and unpleasant (Stewart, 1968; Torkelson et al., 1958). Concentrations between 5000 and 20000 ppm may become life-threatening in some minutes due to the induction of narcosis or to some other mechanism of sudden death (Droz et al., 1982). As regards trichloroethylene and tetrachloroethylene, the odor was unpleasant and strongly irritating when the concentrations exceeded about 600 to 1000 ppm. Lightheadedness, drowsiness, dizziness and nausea have been reported in a few minutes at concentrations ranging from 1000 to 2000 ppm, and complete incoordination and unconsciousness occurred within 30 minutes (Dornette and Jones, 1973; Droz et al., 1982; Torkelson and Rowe, 1982).

Irritation of the eyes, nose and throat are usual symptoms of exposure to high vapour concentrations of alcohols, acetates and ketones. Irritation of the eyes, nose and throat to acetone and methyl ethyl ketone were observed at 500 - 1000 ppm, and 300 - 500 ppm, respectively. Limb weakness, headache, dizziness and lightheadedness have been decribed in accidental exposures to acetone, and the victims recovered fully from an incident in which the concentration exceeded 12 000 ppm (Ross, 1973). Methyl ethyl ketone is narcotic in excess of 8000 ppm. Both ketones are extremely irritating in narcotic concentrations. Similarly, irritation has been reported from an exposure to about 400 ppm of ethyl acetate with no accompanying narcotic symptoms. Among the alcohols isopropanol caused irritation at 400 ppm and higher concentrations, and 1-butanol was claimed to be irritating at 25 - 50 ppm. Although both alcohols affect the central nervous system at very high concentrations, in actual work situations acute inhalational poisonings are unlikely (Nelson et al., 1946, Rowe and McCollister, 1982).

MECHANISMS AND DETERMINANTS OF ACUTE EFFECT

The evaporation rates of organic solvents may vary several hundred-fold. Solvents which evaporate more easily are generally regarded more hazardous than solvents with relatively low evaporation rates. On that basis hazard indices have been developed which take into account both the volatility and the toxicity (as expressed in a hygienic standard) of the solvent. On the other hand it has been shown that the relative potencies of solvents to cause, for example, acute central nervous system depression varies within a wide range.

Clark and Tinston (1982) recently determined the relative potencies of a selection of halogenated and unsubstituted hydrocarbons to elicit powerful acute effects in the central nervous system of the rat and the heart of dogs. The solvents examined caused, after a 10-minute inhalation, the desired level of depression or stimulation of the rat CNS at ambient concentrations ranging from 0.4 to more than 80 vol %, and cardiac sensitization in dogs at 5 minutes over a nearly identical concentration range. They observed that the potencies were inversely related to the vapour pressures of the solvents concerned. In fact the solvents proved nearly equipotent when the results were expressed on a thermodynamic scale, ie. as ratios of the effective partial pressure and the saturated vapour pressure. This has been interpreted to indicate that at high concentrations solvents have a nonspecific physical effect on membranes. The concept of lipid solubility-dependent physical action has long been recognized in anaesthesiology and it has been demonstrated in a variety bioassays (Nielsen and Alarie, 1982).

In practical terms, solvents with high evaporation rates do not necessarily induce acute effects more readily than solvents with low evaporation rates due to the balancing effect of an inverse order of potency. The possibility still remains that some solvents constitute a greater hazard of an acute intoxication than others in actual workplace situations. This could partly be explained by the fact that for poorly soluble solvents effective partial pressures in tissues develop much more readily than for soluble solvents. For example, xylene which is soluble in the blood and tissues and highly soluble in fat induces CNS effects rather slowly because relatively large amounts of the solvent must be transported to the body until an effective partial pressure develops in the brain. Acetone causes narcotic effects even more slowly than xylene. Acetone is a water-soluble solvent and thus it is distributed rather evenly in the whole body. Consequently, it takes a long time to achieve effective partial pressures in the brain

(Bruckner and Petersen, 1981). Conversely, 1,1,1 -trichloroethane would induce narcotic effects fairly rapidly because of its relatively poor solubility in tissues.

CONCLUSIONS

The main acute health hazard of organic solvents relates to their ability to produce narcotic effects. At moderate concentrations the solvents may cause subjective symptoms such as headache, dizziness and drowsiness indicating central nervous system depression. These symptoms are usually transient, and they disappear quickly after termination of the exposure. Respiratory irritation and gastrointestinal symptoms are also common in acute intoxications. At high concentrations anaesthesia and failure of the vital centers of respiration and circulation may develop insidiously or suddenly and lead to death.

Generally the more serious risks are associated with the use of solvents in confined spaces. Young and inexperienced workers seem to be at especially high risk, apparently due to more careless attitudes and a lack of sufficient information and advice. These risks can be greatly reduced with a strict adherence to an appropriate code of practice for entrance and work in confined spaces.

REFERENCES

Bakinson MA, Jones RD (1985). Gassings due to methylene chloride, xylene, toluene and styrene reported to her Majesty´s Factory Inspectorate 1961-80. Br J lnd Med 42: 184-190.

Bass M (1970). Sudden sniffing death. JAMA 12: 2075-79.

Browning E (1965). "Toxicity and Metabolism of Industrial Solvents". Amsterdam: Elsevier Publishing Company.

Bruckner JV, Petersen RG (1981). Evaluation of toluene and acetone inhalant abuse: I. Pharmacology and pharmacodynamics. Toxicol Appl Pharmacol 61: 27-38.

Brugnone F, Perbellini L, Apostoli P, Locatelli M, Mariotto P (1983). Decline of blood and alveolar toluene concentration following two accidental human poisonings. Int Arch Occup Environ Health 53: 157 - 165.

Carpenter CP, Kinkead ER, Geary DL Jr., Sullivan LJ, King JM (1975). Petroleum hydrocarbon toxicity studies. V. Animal and human response to vapors of mixed xylenes. Toxicol Appl Pharmacol 33: 543-558.

Chenoweth MB, Hake CL (1962). The smaller halogenated aliphatic hydrocarbons. Annual Review of Pharmacology 2: 363-398.

Clark DG, Tinston DJ (1982). Acute inhalation toxicity of some halogenated and non-halogenated hydrocarbons. Human Toxicol 1: 239-247.

Dornette WHL, Jones JP (1973). Clinical experiences with 1,1,1-trichloroethane. A preliminary report of 50 anaesthetic administations. Anaesth Analg (Cleveland) 39: 249-250).

Droz PO, Nicole E, Guberan E (1982). Sniffing 1,1,1-trichloroethane simulation of two fatal cases. In Collings AJ, Luxon SG (eds): "Safe Use of Solvents". London: Academic Press, pp 153-159.

Feldman RG, Mayer RM, Taub A (1970). Evidence for peripheral neurotoxic effect of trichloroethylene. Neurology (Minneap.) 20: 599-606.

Folland DS, Schaffner W, Ginn HE, Grofford OB, McMurray DR (1976). Carbon tetrachlororide toxicity potentiated by isopropyl alcohol. JAMA 236: 1853-1856.

Gerdarde HW (1960). "Toxicology and Biochemistry of Aromatic Hydrocarbons". Amsterdam: Elsevier Publishing Company.

Lee TH, Seymour WM (1979). Pneumonitis caused by petrol siphoning. The Lancet 21: 149.

Longley EO, Jones AT, Welch R, Lomaev O (1967). Two acute toluene episodes in merchant ships. Arch Environ Health 14: 481 - 487.

McCarthy TB, Jones RD (1983). Industrial gassing poisonings due to trichloroethylene, perchloroethylene, and 1.1.1.-trichloroethane, 1961-80. Br J lnd Med 40: 450-455.

Morley R, Eccleston DW, Douglas CP, Greville WEJ, Scott DJ, Anderson J (1970). Xylene poisoning--A report on one fatal case and two cases of recovery after prolonged unconsciousness. Brit Med J 3: 442 - 443.

Nelson KW, Ege JF Jr, Ross M, Woodman LE, Silverman L (1946). Sensory response to certain industrial solvent vapors. J Ind Hyg Toxicol 25: 282-85.

Nielsen GD, Alarie Y (1982). Sensory irritation, pulmonary irritation, and respiratory stimulation by airborne benzene and alkyl benzenes: Prediction of safe industrial exposure levels and correlation with their thermodynamic properties. Toxicol Appl Pharmacol 65: 459-477.

Nomiyama K, Nomiyama H (1978). Three fatal cases of thinner-sniffing, and experimental exposure to toluene in human and animals. Int Arch Occup Environ Health 41: 55-64.

Oettingen WF von, Neal PA, Donahue DD (1942). The toxicity and potential dangers of toluene. J Amer Med Ass 118: 579-584.

Ross DS (1973). Acute acetone intoxication involving eight male workers. Ann Occup Hyg 16: 73-75.

Rowe VK, McCollister SB (1982). Alchols. In Clayton GD and Clayton FE (eds): "Patty´s Industrial Hygiene and Toxicology", Vol 2 C, 3rd ed New York: John Wiley & Sons, pp 4561-4707.

Stewart RD (1968). The toxicology of 1,1,1 -trichloroethane. Ann Occup Hyg 11: 71-79.

Stewart RD, Hake CL (1976). Paint remover hazard. JAMA 235: 398-401.

Torkelson TR, Rowe VK (1982). Halogenated aliphatic hydrocarbons containing chloride, bromine and iodine. In Clayton GD and Clayton FE (eds): "Patty´s Industrial Hygiene and Toxicology", Vol 2 B, 3rd ed., New York: John Wiley & Sons, pp 3462-3468.

Wang CC, Irons GV (1961). Acute gasoline intoxication. Arch Environ Health 2:6:714-716.

Winslow C. (1927). Summary of the National Safety Council study of benzol poisoning. 9: 61 - 74.

Wolf MA, Rowe VK, McCollister DD, Hollinsworth RL, Dyen F (1956). Toxicologial studies of certain alkylated benzenes and benzene. A.M.A. Arch Ind Health 14: 387.

Safety and Health Aspects of Organic Solvents, pages 133–138

SOLVENT DERMATITIS

Klaus E. Andersen

Dermatology Clinic, Algade 33, DK-4000
Roskilde, Denmark

INTRODUCTION

Solvents are estimated to be responsible for 6-20% of cases of occupational dermatitis (Schwartz et al., 1957; Adams 1983). Solvents alter the chemical and physical nature of the epidermal barrier, remove the lipid film on the surface and thus diminish the protective capability of the skin (Wahlberg and Boman, 1978; Kronevi et al., 1981; Boman et al., 1982; Wahlberg, 1984 a, b). The destructive effects of the solvents increase the risk of dermatitis - not only from the solvents themselves - but also from irritants and allergens in the environment.

CONTACT DERMATITIS

Contact dermatitis is an inflammatory skin disease caused by external chemicals. The terms dermatitis and eczema are used synonymously. Based on etiology, contact dermatitis can be divided into 5 types: allergic, irritant, contact urticaria, phototoxic and photoallergic dermatitis (Cronin, 1980; Adams, 1983; Marzulli and Maibach, 1983). The clinical picture of all types is extremely variable; frequently they cannot be separated by history and clinical picture alone. Combinations of erythema, edema or induration, papules, vesicles or bullae, scaling and fissuring and excoriations characterize the clinical picture. The patients complain of itching, burning and pain. Clinically they may be divided into acute, subacute or chronic forms. A meticulous history of the patients' daily activities and environmental exposure is required for the classification of contact dermatitis.

ALLERGIC CONTACT DERMATITIS

Allergic contact dermatitis is a T-cell mediated skin disorder diagnosed by an inflammatory reaction to a non-irritant patch test with a suspected allergen. The test is occluded for 48 hours and subsequently evaluated. Patch testing is routinely performed by applying a standard series of the most frequently occurring contact allergens. The tray is regularly updated. The standard series reveals about 80% of contact sensitivities. The remaining ones are found by supplementary tests with other allergens. The currently used standard series contain metals, topical drugs, rubber accelerators, balsams, formaldehyde, epoxy and a few others. Cronin (1980) details the technique and the pitfalls of patch testing.

SENSITIZING SOLVENTS

Turpentine is an oleoresin obtained from pine trees. Oil of turpentine contains a variable mixture of terpenes and their sensitizing autooxidation products - hydroperoxides (Pirilä and Siltanen, 1958). It is a former thinner in paints and varnishes - now replaced by non-sensitizing alternatives like mineral turpentine (white spirit); subsequently contact allergy to the compound has become rare.

Various alcohols and glycols are used as solvents and may occasionally cause allergic contact dermatitis (Cronin, 1980; Adams, 1983). Propylene glycol is a good solvent widely used in foods, drugs, cosmetics and in industry. It may produce eczematous skin reactions of toxic and, more rarely, of allergic nature (Andersen and Storrs, 1982). Positive patch test reactions to propylene glycol are difficult to interpret. Allergic reactions may be confirmed by a clear clinical relevance, repeated local skin provocation (usage test), or oral provocation (Hannuksela and Förström, 1978). Dioxane (diethylene ether) is a degreasing solvent which sensitized a worker who used it to clean metal parts (Fregert, 1974).

Formaldehyde may in some cases be regarded as a solvent (Adams, 1983), and is a ubiquitous and potent allergen included in the standard patch test series. Exposure may occur inadvertently as more than 80 trade names and synonyms are used in the marketing of formaldehyde releasing compounds (Fiedler, 1983).

IRRITANT DERMATITIS

Almost all solvents are more or less potent irritants. Our knowlegde about pathogenesis and mechanisms of irritant dermatitis is limited. Acute dermatitis is elicited by strong irritants after a single or a few applications. This is easily diagnosed by history, and often involves occupational accidents. The clinical appearance varies depending on the substances from deep red corrosions to dermatitis indistinguishable from acute allergic contact dermatitis. Much more frequent is the cumulative irritant dermatitis, where repeated insults by low grade irritants over a long period are required. Dryness and cracking of the skin are often the initial characteristics. Redness, scaling, papules, vesicles and a gradual thickening of the skin may supervene.

The etiological factors are complex and usually a combination of irritants are involved. The initial dermatitis may be caused by a strong irritant as a solvent and then sustained by soaps and detergents. An important clinical feature is that the skin all over the body reacts as one organ: An acute irritant dermatitis of the hands lowers the threshold for irritant reactions on the back (Björnberg, 1968). This should be recognized when patch testing is performed as false positive reactions may develop (Mitchell and Maibach, 1982).

The irritant hand dermatitis frequently starts in moist areas difficult to rinse and dry as under rings and in the interstices. It may spread to the dorsum of the hand where the skin is thinner and less resistant than in the palms. Irritant contact dermatitis may be long lasting, if not treated soon after onset. Even when it appears to be healed, the protective capacity of the skin is still impaired for several weeks to several months. Further, irritant contact dermatitis may be a precursor of the development of an allergic contact dermatitis.

There appears to be a predisposition to irritant hand dermatitis. Individuals who had atopic dermatitis during childhood are more prone to get hand dermatitis when employed in certain occupations with wet work and dirt (Lammintausta, 1981; Rystedt 1985).

However, Björnberg (1974) examined skin reactivity to a range of irritants in individuals with healed hand eczema and in matched controls. The two groups showed about the same intensity of the skin reactions to the irritants. Thus it was not possible to predict an individual's predisposition to eczema by testing the normal skin with irritants. In this study patients with atopic dermatitis and vesicular hand dermatitis were excluded.

There is no objective test to diagnose an irritant dermatitis. Because of the clinical similarity to allergic contact dermatitis it is important that all cases are patch tested.

The irritant capability of solvents varies: The poorly absorbed solvents cause more skin damage and low systemic toxicity in contrast to the solvents which penetrate easily and cause systemic toxicity without causing considerable skin damage. The saturated hydrocarbons are more irritating than the solvents from the aromatic series (Schwartz et al.,1957). However, many solvents cause considerable microscopical changes within minutes of exposure without causing visible (macroscopical) changes (Kronevi et al.,1981).

Trichlorethylene caused severe generalized dermatitis and systemic toxicity in 4 workers (Bauer and Rabens, 1974). Exposure could be monitored by the presence of trichloro-acetic acid in the urine. Further, alcohol ingestion combined with exposure to trichlorethylene vapor caused transient facial erythema in 7 volunteers (Stewart et al., 1974).

SKIN SCLEROSIS

There are a few reports about other skin disorders related to solvent exposure: Occupational scleroderma, which is a serious sclerosis (stiffness and thickening) of the connective tissue in the skin may be caused by vinyl chloride and perchloroethylene (Sparrow, 1977). Naphta, n-hexane, and hexachloroethane are suspect sclerosis causing solvents in animal experiments using guinea pigs (Yamakage and Ishikawa, 1982).

TREATMENT AND PREVENTION

The treatment of solvent dermatitis is the same as for other types of contact dermatitis: topical corticosteroids and skin care, replacement of the provoking factors with alternative products and not the least important: information and protection.

Barrier creams is one possibility. The problem is that barrier creams are poor barriers. It is very difficult to find good evidence for their efficiency (Boman et al, 1982). The major benefit from their use may be that they make it easier to clean the skin after work. The ideal attributes of barrier creams are the following: impermeable to the hazardous substance, non-irritating and non-sensitizing, easy to apply, it should not interfere with the work, have a persistent effect, easy to remove after work, and non-perishable during storage.

Protection gloves are another possibility for skin protection. Their effect depends on the composition, thickness, surface texture, method of manufacture and interaction between glove material and the chemical. Detailed knowledge of the chemical exposure is required for the optimal choice of glove material (Williams, 1979). Gloves may also be destroyed by the solvents.

REFERENCES

Adams RM (1983). " Occupational skin disease." New York: Grune & Stratton.

Andersen KE, Storrs FJ (1982). Hautreizungen durch Propylenglykol. Hautarzt 33:12-14.

Bauer M, Rabens SF (1974). Cutaneous manifestations of trichloroethylene toxicity. Arch Dermatol 110:886-890.

Björnberg A (1974). Skin reactions to primary irritants and predisposition to eczema. Br J Dermatol 91:425-427.

Björnberg A (1968). " Skin reactions to primary irritants in patients with hand eczema." Göteborg: Oscar Isacsons Tryckeri AB.

Boman A, Wahlberg JE, Johansson G (1982). A method for the study of the effect of barrier creams and protective gloves on the percutaneous absorption of solvents. Dermatologica 164:157-160.

Cronin E (1980). " Contact dermatitis." Edinburgh: Churchill Livingstone.

Fiedler HP (1983). Formaldehyd - Formaldehyd-abspalter. Dermatosen 31:187-189.

Fregert S (1974). Allergic contact dermatitis from dioxane in a solvent for cleaning metal parts. Contact Dermatitis Newsletter 15:438.

Hannuksela M, Förström L (1978). Reactions to peroral propylene glycol. Contact Dermatitis 4:41-45.

Kronevi T, Wahlberg JE, Holmberg B (1981). Skin pathology following epicutaneous exposure to seven organic solvents. Int J Tiss Reac 3:21-30.

Lammintausta K, Kalimo K (1981). Atopy and hand dermatitis in hospital wet work. Contact Dermatitis 7:301-308.

Marzulli FN, Maibach HI (1983) " Dermatotoxicology." ed. 2, Washington: Hemisphere Publishing Corporation.

Mitchell JC, Maibach HI (1982). The angry skin syndrome - the excited skin syndrome. Semin Dermatol 1:9-13.

Pirilä V, Siltanen E (1958). On the chemical nature of turpentine III. Dermatologica 117:1-8.

Rystedt I (1985). Work-related hand eczema in atopics. Contact Dermatitis 12:164-171.

Schwartz L, Tulipan L, Birmingham DJ (1957). " Occupational diseases of the skin." ed.3, Philadelphia: Lea & Febiger.

Sparrow GP (1977). A connective tissue disorder similar to vinyl chloride disease in a patient exposed to perchlorethylene. Clin Exp Dermatol 2:17-22.

Stewart RD, Hake CL, Peterson JE (1974). "Degreasers' flush" Arch Environ Health 29:1-5.

Wahlberg JE (1984a). Edema inducing effects of solvents following topical administration. Derm Beruf Umwelt 32:91-94.

Wahlberg JE (1984b). Erythema inducing effects of solvents following epicutaneous administration to man - studies by laser Doppler flowmetry. Scand J Work Environ Health 10:159-162.

Wahlberg JE, Boman A (1978). Comparative percutaneous toxicity of ten industrial solvents in the guinea pig. Scand J Work Environ Health 5:345-351.

Williams JR (1979). Permeation of glove materials by physiologically harmful chemicals. Am Industr Hyg Ass J 40:877-882.

Yamakage A, Ishikawa H (1982). Generalized morphea-like scleroderma occurring in people exposed to organic solvents. Dermatologica 165:186-193.

Safety and Health Aspects of Organic Solvents, pages 139–154

SOLVENTS AND THE LIVER

Matti Klockars, M.D.

Institute of Occupational Health
Haartmaninkatu 1, 00290 Helsinki, Finland

Introduction

Many chemicals are known to be toxic to the liver. However, the relative importance of the various toxins causing hepatic injury in man has changed considerably over the years. In Scandinavia, this is partly because many toxic compounds have been more or less eliminated from the working environment, or else they are handled under reasonably strict hygienic control.

Epidemiology

In order to establish a connection between solvents and hepatic injury epidemiologic studies require large cohorts of workers. From the epidemiological point of view such studies are difficult to carry out. Some of the epidemiological studies that have positively linked solvents and liver disease have failed to provide sufficient information on levels of exposure or to apply strict enough controls to fulfill the criteria of valid epidemiological research (Kurppa and Vainio, 1983). Good epidemiological research should at least fulfill these two criteria of study design:

1. A validly selected and defined "exposed" group
2. A carefully selected and proper "control" group

The investigators should be free of a priori assumptions of a cause-effect relationship between solvents and liver injury before the start of the study. In analyzing biochemical variables, reference values of the routine clinical laboratory should not and cannot be used (Kurppa and Vainio, 1983). Great difficulties may occur when matching the exposed population to the control population for factors such as alcohol, drug intake, previous liver disease and life-style. However, epidemiological studies to identify hepatotoxins should be given high priority.

Classification of hepatotoxic agents

Clinical and experimental observations have led to general agreement about two main groups of substances that produce hepatic injury (Zimmerman and Maddrey, 1982).

1. Intrinsic, predictable or true hepatotoxins. Hepatotoxicity is a fundamental property of the agent, and most exposed individuals are susceptible to them.

2. Idiosyncratic, non-predictable hepatotoxins. These hepatotoxins produce hepatic injury only in unusually susceptible individuals

Hepatic injury due to host idiosyncracy is by definition unpredictable and occurs in a small number of recipients. This may be a manifestation of the solvent taking an aberrant metabolic pathway in an susceptible worker. The basis for this metabolic aberration may be genetic or it may be acquired as the result of exposure to other agents which induce "new" metabolic pathways.

There are two types of intrinsic hepatotoxins (Zimmerman and Maddrey, 1982):

a) Direct hepatotoxins

These agents, or their metabolites, injure the hepatocyte and its organelles by a direct physico-chemical effect that may culminate in necrosis of liver cells. Thus the metabolic disorder is secondary to structural damage.

b) Indirect hepatotoxins

These agents are more selective and discriminative.

They produce hepatic injury mainly by interference with specific metabolic pathways. The structural injury is thus secondary to a metabolic lesion. Ethanol and vinyl chloride are examples of indirect hepatotoxins.

Clinical evaluation of the association between solvents and liver injury

For a clinician the problem of occupational exposure to solvents and liver disease is generally not that of the etiologic analysis of an acute massive toxic hepatic necrosis, but one is usually faced with a worker who has abnormal liver enzymes on a routine blood test, and who works with a variety of chemicals and possibly occasionally drinks alcohol. The question is, to what extent the observed changes in liver enzyme levels could be related to the exposure to solvents?

The diagnostic procedure usually includes the following questions.

i) Does the patient have symptoms and signs of hepatic impairment?
ii) Does the patient have a history of exposure to chemicals?
iii) Is it possible to exclude other factors giving the same symptoms and signs?

In clinical practice the diagnosis of hepatic impairment is usually based on the analysis of biochemical variables and histologic findings. It is important to realize that hepatic injuries produced by chemicals do not have unique clinical, laboratory or morphologic features. The patient's history of occupational exposure to solvents should be verified both qualitatively and quantitatively.

When selecting tests for liver function for use in occupational health practice one has to pay attention to the following criteria:

SPECIFICITY - the more specific a test with regard to liver function the higher its diagnostic value

SENSITIVITY - the more sensitive the test, the earlier the stage at which it can detect a change in liver function

Because of the diversity of hepatic metabolism and function, a variety of biochemical tests have been used to study hepatic injury (Table 1). However, none of the blood tests used routinely to assess liver function is ideal.

Table 1. A brief summary of laboratory tests on serum samples used to evaluate various types of liver diseases.

Hepatocellular injury	Hepatocanalicular injury
Aspartate aminotransferase	Alkaline phosphatase
Alanine aminotransferase	Gamma glutamyl transferase
Ornithine-carbamoyl transferase	Leucine aminopeptidase
Glutamate dehydrogenase	5'Nucleotidase
Isocitrate dehydrogenase	
Hepatic function	**Hepatic fibrosis**
Albumin synthesis	Prolyl hydroxylase
Ammonia production	Lysyl hydroxylase
Bilirubin concentration	Collagen galactosyl transferase
Sulfobromophthalein (BSP) clearance	Collagen glucosyl transferase
Antipyrine clearance	
Galactose clearance	
Gamma glutamyl transferase	
Bile acid concentrations	

Immunological hepatic injury and chronic hepatitis

Hepatitis virus (A and B) antigens and antibodies
Immunoglobulin concentrations
Complement components (C3 and C4)
Antinuclear antibodies
Smooth muscle antibodies
Mitochondrial antibodies
Alfa-1-antitrypsin phenotype and concentration
Ceruloplasmin concentration

Other chemical tests (liver biopsy samples)

Depression of hepatic glucose-6-phosphatase activity as a measure of microsomal injury
Hepatic cytochrome P-450 activity.

Measurement of aspartate and alanine transferases, alkaline phosphatase and gamma glutamyl transferase are requested as an aid to diagnose and monitor possible chemically-induced hepatic injury. When interpreting minor changes in enzyme activity, the significance and importance of physiological variations or variations attributed to age, weight, alcohol consumption or drug intake must be carefully considered (Siest et al, 1975).

The microsomal enzyme system represents a multitude of enzymes primarily confined to the smooth endoplasmic reticulum of hepatocytes. These enzymes are responsible for the metabolism of a wide variety of exogenous compounds, including industrial chemicals. The activity of the hepatic microsomal enzyme system can be changed by many occupationally used chemical agents at concentrations lower than those causing manifest clinical hepatotoxicity

The most widely used in vivo noninvasive assessment of hepatic microsomal enzyme activity in man is the antipyrine clearance test.

Table 2. Examples of antipyrine metabolism among persons with occupational exposure to chemicals (Dössing, 1984).

Chemicals	Job category	Antipyrine metabolism
Chlorinated hydrocarbons	Aerosol spraying	increased
Polychlorinated diphenyls	Manufacturing capacitors	increased
Petrol	Petrol station attendance	increased
Solvents, metals, etc.	Spray painting	decreased
Pesticides, solvents	Spraying	increased
Styrene (50 ppm)	Manufacturing polyester plant	unchanged
Solvents	Printing	unchanged
Toluene (100 ppm)	Printing	unchanged

Table 2 shows changes that have been observed among persons with occupational exposure to chemicals. The main question in analyzing the occupational effects of liver function is whether the recorded, reversible changes in microsomal enzyme activity are purely adaptive and harmless or potentially harmful? The answer is that we do not know. However, according to Dössing (1984): "It seems justifiable to regard changes in microsomal enzyme activity as biological changes with potentially harmful consequences to the human organism".

Chemical injury of the liver

Chemical hepatic injury has been encountered in a variety of circumstances. A list of workplace chemicals which may produce hepatic effects in humans is given Table 3.

Table 3. Solvents found or suspected to have hepatotoxic potential in man.

IN COMMON USE	USED FOR SPECIFIC PURPOSES
Alcohols and derivatives	Aliphatic halogenated hydrocarbons
Ethyl alcohol	Carbon tetrachloride
Methyl alcohol	Chloroform
Ethylene glycol ethers and derivatives	1,1 dichloroethane
	Tetrachloroethane
Esters	Nitriles
Ethyl acetate	Acetonitrile
N-butyl acetate	Acrylonitrile
Aliphatic halogenated hydrocarbons	Aromatic hydrocarbons
Methylene chloride	Diphenyl
Tetrachloroethylene	
1,1,1 trichloroethane	Miscellaneous chemicals
Trichloroethylene	Carbon disulfide
	Cyclohexane
Aromatic hydrocarbons	Dioxane
Styrene	Ethers and epoxy compounds
Ethyl benzene	Ethyl ether
Toluene	
Xylene	
Miscellaneous organic	
Nitrogen compounds	
N,N-dimethylformamide	

Although potential hepatic effects cannot be predicted from the chemical structure of an agent, some general features have been suggested to indicate a toxic potential. For hydrocarbons, the hepatic effect increases with the number of halogen atoms fixed on the molecules, is more marked with unsaturated subtances than with saturated derivatives, and is more marked with asymmetric molecules (Efthymiou, 1979).

The current knowledge of chemical hepatotoxicity has been derived mainly from medical case reports describing suicidal or accidental ingestion of chemicals or environmental exposure of groups of individuals. Very little, however, is known about the effects on man by the chemicals at low dose, long-term exposure, or in combined exposures with other chemical agents, or drugs.

Carbon tetrachloride and chloroform

In 1936, Cameron and Karunaratne reported that repeated administration of CCl_4 to rats resulted in cirrhosis of the liver. CCl_4 is hepatotoxic in man, although recent cases of such intoxication are extremely rare.

Chloroform is also a potent hepatotoxin, causing both fatty liver and, at higher levels of exposure, centrilobular necrosis.

Trichloroethylene

Acute trichloroethylene exposure in industrial accidents or during "sniffing" has resulted in acute neurologic, renal and hepatic toxicity indistinguishable from that seen with carbon tetrachloride.

Case reports from the 1960s of chronic industrial trichloroethylene inhalation with hepatic damage have been published (Priest and Horn, 1965; Smith, 1966). However, prolonged exposure of a variety of experimental animals to trichloroethylene via inhalation has produced neither significant dysfunction nor abnormal hepatic histology (Prendergast et al, 1967).

Three teenagers with a history of drug abuse developed acute hepatic injury after inhaling trichloroethylene containing cleaning fluid. Hepatic biopsies of two of these patients showed acute centrilobular necrosis (Baerg and Kimbery, 1970).

1,1,1,-trichloroethane (methylchloroform)

Acute 1,1,1-trichloroethane (TCE) exposure has been associated with acute hepatotoxicity in rare case reports of human subjects (Stewart and Andrews, 1966; Nathan and Toseland, 1979). In a case report by Halevy et al. 1980, acute overexposure to TCE vapour resulted in neurological symptoms and the development of short term liver damage two days after the accident. Moderately elevated liver enzyme values were recorded, as well as abnormal renal function values. A percutaneous liver biopsy showed inflammatory cells, including eosinophils, in the periportal areas and cholestasis. At the electron microscopic level, proliferation of peroxisomes and smooth endoplasmic reticulum was seen. The liver function returned to normal. Interestingly some features in this case, namely urticarial rash and eosinophilic infiltration in the liver, may point to an individual hypersensitivity reaction to TCE in this patient.

In an epidemiologic study of workers in a textile plant the health effects of TCE were investigated. In the exposure groups of 1-250 ppm, enzyme studies did not reveal any association between increased exposure and hepatic dysfunction (Kramer et al, 1978).

A case report describes the development of liver cirrhosis and portal hypertension after repeated bouts of acute hepatotoxicity caused by trichloroethylene and a final episode of 1,1,1-trichloroethane exposure (Thiele et al, 1982).

Toluene

Toluene has occasionally caused liver injuries. Heavy chronic exposure, such as occurs in sniffing or accidentally in occupational settings, has resulted in liver damage, e.g., fatty infiltration and necrosis of single cells (Gattner and May, 1963).

However, in two human cases of acute intoxications caused by occupational exposure to toluene, which resulted in unconsciousness, there was no impairment of liver function during a complete recovery from the accident (Brugnone and Perbellini, 1985).

In a study by Tähti and co-workers (1981) on 78 workers occupationally exposed to toluene (20-200 ppm), no significant differences were observed in the serum concentration of aminotransferases, compared with a non-exposed group.

Xylene

Xylene is not a common cause of liver injury. Even in acute severe xylene intoxications only moderate increases in serum aminotransferases have been reported, indicating low hepatotoxicity of xylene. In the description of three severe, including one fatal, cases of xylene poisoning after prolonged inhalation of paint fumes there was only a slight elevation of aminotransferases in the two surviving patients during recovery (Morley et al, 1970)

Styrene

Acute or chronic styrene exposure has generally not produced any observable liver changes in man.

Styrene, at concentrations in use in a Finnish laminating plant did not have any significant effect compared with a referent population, on the levels in serum of aspartate aminotransferase, alanine aminotransferase, gamma glutamyl transferase, cholic acid and deoxycholic acid (Härkönen et al, 1984).

Raised concentrations of conjugated bile acids have been observed among workers exposed to styrene (Edling and Tageson, 1984). It is possible that the determination of serum bile acids, cholic acid and chenodeoxycholic acid may be more sensitive and early indicators of hepatic dysfunction.

Methylene chloride

Methylene chloride appears not to be hepatotoxic in man.

Exposure of human volunteers to 200-1000 ppm of methylene chloride for 1-2 hours did not affect liver function (Stewart et al, 1972). Even fatal accidental exposure did not produce any marked liver injury (Moskowitz and Shapiro, 1952).

Isopropanol

Isopropanol given to rats in high repeated doses increased the concentration of triglycerides of the liver (Beauge et al, 1972). There are no reports of a similar effect in humans.

1-Butanol

There are no reports that pure butanol causes any liver injury in humans.

Butanone

Butanone is extensively used as an industrial solvent and is found in some commercial and household glues. A case report of self-poisoning by ingestion of butanone reports that hepatic function remained normal during recovery of a comatose patient with metabolic acidosis (Kopelman and Kalfayan, 1983).

Carbon disulfide

Exposure to carbon disulfide has caused fatty degeneration and haemorrhages of the liver in animals. This is particularly the case if the animals have been pretreated with phenobarbital to stimulate the microsomal enzyme activity of the liver. The effect is further potentiated by starvation (Magos and Butler, 1972).

Workers hospitalized for chronic CS_2 poisoning (with extremely high exposure levels) showed some functional disturbances and fatty degeneration of hepatocytes, but no necrotic changes of liver cells (WHO, 1979). Mixed function oxidases have been shown to be inhibited after only a single 6-hour exposure to 10 ppm of carbon disulfide.

Dioxane

There are a few case reports (6 cases) of hepatic injury caused by dioxane in the 1930s and 1950s (Barber, 1934; Johnstone, 1959).

Halothane

From the chemical and toxicological point of view, halothane, (2-bromo-2-chloro-1,1,1-trifluoroethane), can be regarded as an organic solvent. The hepatotoxicity of this anesthetic solvent has some features that are of interest when considering the pathogenetic mechanism of solvent induced liver injury.

Disturbed hepatic function has been associated with the occupational use of halothane as an anaesthetic among health personnel (Klatskin and Kimberg, 1969, Neuberger and Williams, 1984) and exposure to trace concentrations of halothane among anaesthetists has resulted in enzyme induction as measured with the antipyrene clearance test (Duvaldestin et al, 1981). There has been some evidence that frank liver injury could occur as a manifestation of an idiosyncratic allergic reaction to halothane. Patients with halothane-induced hepatitis have been shown to react to the solvent with lymphocyte transformation (Paronetto and Popper, 1970).

Two surgeons with halothane induced hepatic injury were found to have a circulating antibody which reacted specifically with halothane-altered hepatocyte membrane components (Neuberger et al, 1981). The idiosyncratic nature of halothane induced fulminant hepatic failure may reflect quantitative differences in the rate and route of halothane metabolism and hence in the membrane exposure of the altered antigen. The immune recognition of the new antigen or the intensity of the resulting immune response may also be individual characteristics determining the response to halothane (Vergani et al, 1980).

Solvent mixtures

Although environmental exposure to mixtures is far more common than exposure to single toxicants, only a few controlled studies on the hepatic effects of exposure to solvent mixtures in industry have been reported. Liver enzyme activities in the serum of car painters compared with age matched referents have been studied (Kurppa and Husman, 1982). The painters were exposed to a mixture of solvents including toluene, xylene and other constituents at a concentration of about half the threshold limit value recommended by the American Conference of Governmental Industrial Hygienists in 1981. The activities of aspartate aminotransferase, alanine aminotransferase, ornithine carbamoyl transferase and gamma glutamyl transferase did not differ between the exposed and the unexposed groups.

Potentiation of solvent toxicity by ethanol

Carbon tetrachloride, trichloroethylene and methylene chloride

The consumption of ethanol prior to exposure to CCl_4 vapours results in increased toxicity. In experimental animals enhanced toxicity is most apparent when ingestion of ethanol precedes exposure to CCl_4 by 16 to 18 hours. (Cornish and Adefuin, 1966). When one makes an extrapolation of the data obtained from animal experiments to possible human conditions, one may expect a moderate consumption of ethanol (40 to 80 g daily) to increase the hepatotoxic effects of CCl_4. Thus, persons who consume alcohol should avoid any exposure to carbon tetrachloride (Hills and Venable, 1962).

Alcohol intolerance after occupational exposure to trichloroethylene has been reported (Sbertoli and Brambilla, 1962). Trichloroethylene exposure significantly increased the blood half-life of ethanol. Ethanol dehydrogenase not only converts ethanol to acetaldehyde, but also biotransforms the trichloroethylene metabolite chloral hydrate to trichloroethanol. Competitive inhibition of trichloroethylene metabolism results in higher blood levels of trichloroethylene. Thus concentrations of the trichloroethylene end metabolite, trichloroacetic acid, were in ethanol-treated animals 1/100 of that seen in control animals (White and Carlson, 1981).

Because trichloroethylene treatment decreases the rate of elimination of ethanol from blood, a concomitant administration of trichloroethylene and ethanol may result in a mutual inhibition of their metabolism. Oxidation of trichloroethylene to trichloroacetic acid did not occur as long as ethanol was detectable in the blood (Müller et al, 1975). Because these effects have been shown with low ethanol concentrations, this is a condition which may arise in humans working with trichloroethylene.

There is some indication that for one-off exposures or for very short-term exposure to both methylene chloride and ethanol, the interaction is antagonistic. However, over time, the hepatotoxic effects of methylene chloride appear to be aggravated by ethanol (Balmer et al, 1976)

In addition to ethanol, the effects of CCl_4 are potentiated by isopropanol, acetone, 2-butanol, iron and vanadate. Effects of chloroform are potentiated by ethanol, isopropanol, acetone, n-hexane, methyl n-butyl ketone, 2,5-hexanedione and chlordecone (a ketonic insecticide).

Conclusion

It is possible that there are many hitherto unrecognized hepatotoxic compounds while little is known about the potentially dangerous interaction of different chemicals within the human body. We have little information about the potential toxic effects of long-term exposure to low doses of many of the newer chemical compounds used and produced in industrial processes. We also need simple non-invasive tests of liver function since those currently available are clearly inadequate for early detection of toxic liver damage.

According to the Swedish author Dr. Olav Axelsson (1983), it seems at the moment prudent to maintain an open mind about the possibility that solvents might be etiologically involved in various types of liver damage and other adverse health effects. Axelsson continues by stating that "In view of current knowledge and the difficulties involved in a more definite evaluation of the relationships between solvent exposure and liver damage, it seems natural at this time to take a somewhat critical attitude as to whether there is a causal link or not between exposures to commonly used solvents and liver damage".

REFERENCES

Axelson O. Solvents and the liver. Eur J Clin Invest, 13:109-111, 1983

Baerg RD, Kimberg DV. Centrilobular hepatic necrosis and acute renal failure in "solvent sniffers". Ann Int Med 73:713-720, 1970

Balmer MF, Smith FA, Leach LJ. Yuile CL. Effects in the liver of methylene chloride inhaled alone and with ethyl alcohol. Am Ind Hyg Ass J 37:345-352, 1976

Barber H. Haemorrhagic nephritis and necrosis of the liver from dioxine poisoning. Guy's Hosp Rep 84:267-280, 1934

Beauge F, Clement M, Guidicelli Y, Nordmann R, Nordmann J. Effect of isopropanol on palmitate 1-^{14}C incorporation in hepatic triglycerides and phospholipids in the rat. C R Acad Sci (D) (Paris) 275:3005-3008, 1972

Brugnone F, Perbellini L. Toluene coma and liver function. Scand J Work Environ Health 11:55, 1985

Cameron GR, Karunaratne WAE. Carbon tetrachloride cirrhosis in relation to liver regeneration. J Pathol Bacterial 42:1-21, 1936

Cornish HH, Adefuin J, Ethanol potentiation of halogenated aliphatic solvent toxicity. Am Ind Hyg J 27:57-61, 1966

DeFalque RJ. Pharmacology and toxicology of trichloroethylene. A critical review of the world literature. Clin Pharmacol Ther 2:665-675, 1960

Duvaldestin P, Mazze RI, Nivoche Y, Demonts JM. Occupational exposure to halothane results in enzyme induction in anesthetists. Anesthesiology 54:57-60, 1981

Dössing M. Noninvasive assessment of microsomal enzyme activity in occupational medicine: Present state of knowledge and future perspectives. Int Arch Occup Environ Health 53:205-218, 1984

Edling C, Tagesson C. Raised serum bile acid concentrations after occupational exposure to styrene: a possible sign of hepatotoxicity? Br J Ind Med 41:257-259, 1984

Efthymiou ML. Hepatotoxic industrial solvents. Med Chir Dig 8:381-382, 1979

Gattner H, May G. Panmyelopathic durch Toluoleinwirkung. Zentralbl Arbeitsmed Arbeitsschutz 13:156-157, 1963

Halevy J, Pitlik S, Rosenfeld J. 1,1,1,-trichloroethane intoxication: A case report with transient liver and renal damage. Review of the literature. Clin Toxicol 16:467-472, 1980

Hills BW, Venable HL. The interaction of ethyl alcohol and industrial chemicals. Am J Ind Med 3:321-333, 1982

Härkönen H, Lehtniemi A, Aitio A. Styrene exposure and the liver. Scand J work environ health 10:59-61, 1984

Johnstone RT. Death due to dioxane. AMA Arch Ind Health 20:445-447, 1959

Klatskin G, Kimberg DV. Recurrent hepatitis attributable to halothane sensitization in an anaesthetist. New Engl J Med 280:515-522, 1969

Kopelman PG, Kalfayan PY. Severe metabolic acidosis after ingestion of butanone. Br Med J 186:21-22, 1983

Kramer CG, Ott GM, Fulkerson JE, Hicks N, Imbus, HR. Health of workers exposed to 1,1,1,-trichloroethane: a matched-pair study. Arch Environ Health 33:331-342, 1978.

Kurppa K, Husman K. Car painters' exposure to a mixture of organic solvents. Serum activities of liver enzymes. Scand J Work Environ Health 8:137-140, 1982

Kurppa K, Vainio H. Study design, liver disease and house painters' exposure to organic solvents. Eur J CLin Invest 13:113-114, 1983

Magos L, Butler WH. Effect of phenobarbitone and starvation on hepatotoxicity in rats exposed to carbon disulphide vapour. Brit J Industr Med, 29:95-98, 1972

Morley R, Eccleston DW, Douglas CP, Greville WEJ. Scot DJ, Anderson J. Xylene poisoning: a report on one fatal case and two cases of recovery after prolonged unconsciousness. Br Med J 3:442-443, 1970

Moskowitz S, Shapiro H. Fatal exposure to methylene chloride vapor. Arch Ind Hyg Occup Med 6:116-125, 1952

Müller G, Spassowski M, Henschler D. Metabolism of trichloroethylen in man. III. Interaction of trichloro-ethylene and ethanol. Arch Toxicol 33:173-189, 1975

Nathan AW, Toseland PA. Goodpasture's syndrome and trichloroethane intoxication. Br J Clin Pharmacol 8:284-286, 1979

Neuberger J, Vergani D, Mieli-Vergani G, Davis M, Williams R. Hepatic damage after exposure to halothane in medical personnel. Br J Aneasth 53:1171-1177, 1981

Neuberger J, Williams R. Halothane anaesthesia and liver damage. Br Med J 289:1136-1139, 1984

Paronetto F, Popper H. Lymphocyte stimulation induced by halothane in patients with hepatitis following exposure to halothane. New Engl J Med 283:277-280, 1970

Prendergast JA, Jones RA, Jenkins LJ Jr. Effects on experimental animals of long-term inhalation of trichloroethylene, carbon tetrachloride, 1,1,1-trichloroethane, dichlorodifluoromethane, and 1,1-dichloroethylene. Toxicol Appl Pharmacol 10:270-289, 1967

Priest RJ, Horn RC. Trichloroethylene intoxication - a case of acute hepatic necrosis possibly due to this agent. Arch Environ Health 11:361-365, 1965

Sbertoli C, Brambilla G. Three cases of intolerance a alcohol as sole symptom of trichloroethylene poisoning. Med Lav 53:353, 1962

Siest G, Schiele F, Galteau MM, Panek E, Steinmetz J, Fagnani F, Gueguen R. Aspartate aminotransferase and alanine aminotransferase activities in plasma: Statistical distributions, individual variations, and reference values. Clin Chem 21:1077-1087, 1975

Smith GF. Trichloroethylene: a review. Brit J Industr Med 23:249-262, 1966

Stewart RD, Andrews JT. Acute intoxication with methylchloroform. JAMA 195:904-906, 1966

Stewart RD, Fisher TN, Hosko MJ, Peterson JE, Baretta ED, Dodd HC. Experimental human exposure to methylene chloride. Arch Environ Health 25:342-348, 1972

Thiele DL, Eigenbrodt EH, Ware AJ. Cirrhosis after repeated trichlorothylene and 1,1,1-trichloroethane exposure. Gastroenterol 83:926-929, 1982

Tähti H, Kärkkäinen S, Pyykkö K, Rintala E, Kataja M, Vapaatalo H. Chronic occupational exposure to toluene. Int Arch Occup Environ Health 848:61-69, 1981.

WHO. Environmental health criteria 10, 1979

Vergani D. Mieli-Vergani G, Alberti A, Neuberger J, Eddleston ALWF, Davis M, Williams R. Antibodies to the surface of halothane-altered rabbit hepatocytes in patients with severe halothan-associated hepatitis. New Engl J Med 303:66-71, 1980

White JF, Carlson GP. Epinephrine-induced cardiac arrhythmias in rabbits exposed to trichloroethylene: potentiation by ethanol. Toxicol Appl Pharmacol 60:466-471, 19817Zimmerman HJ, Maddrey WC. Toxic and drug-induced hepatitis. In: Diseases of the liver. Fifth edition. Ed. Leon Schiff and Eugene R Schiff. The J B Lippincott Company 1982

Safety and Health Aspects of Organic Solvents, pages 155–167

SOLVENTS AND THE KIDNEY

Alf Askergren

The Construction Industry's Organisation for Working Environment, Safety and Health, Sweden.

INTRODUCTION

A short review of the renal structures and functions facilitates the understanding of why the kidney may be an essential target of toxic agents (Bulger RE, 1979; Hook JB, 1980; Ryan GB, 1981):

About 25 % of the cardiac output, which is about 1200 ml blood/min, passes through the kidneys. Each kidney consists of about 1 million functional units called nephrons. Fluid and substances are filtered in the first, upper part of the nephron, the glomerulus. Factors of importance for the result of this filtration are the hydrostatic and osmotic pressure, molecular size and electric charges of molecules and tissue structures.

Approximately 170 liters of filtrate are normally produced in this way during a 24 hour period. When the filtrate passes down the nephron and the collecting ducts, a reabsorbtion and an excretion of both fluid and substances takes place. Finally about 1-2 liters are excreted as urine.

While the filtration is a relatively simple process, the work exerted by the rest of the nephron and the ducts is indeed complicated, involving metabolically and enzymatically very active processes. These include numerous biotransformations, which may result in more biologically active substances being presented to the renal parenchyma.

The reabsorbtion of fluid and substances takes place through cells and membranes lining the tubules and ducts. A

concentration of substances within the cells may result. Filtered substances which are not reabsorbed through the tubular walls, will become highly concentrated during their passage down the nephron.
Another outcome of these operations is a concentration of substances in the interstitial tissues surrounding the lower parts of the nephron.

There are however not only theoretical reasons to believe in a possible nephrotoxic effect by such substances as for instance hydrocarbons (HC). Results of animal studies, reports on human cases and studies of human populations clearly indicate an association between organic solvent exposure and renal disease. Such studies will be shortly reviewed here.

ANIMAL STUDIES

Organic solvents´ toxic effects on the kidneys of different species, especially mouse and rat, have been known for decades. Members of all the main groups of solvents have been studied and found toxic to various extent. The damage may be acute and/or chronic and situated in varous parts of the nephron, especially the tubules. It may present itself as tubular degeneration with regenerative epithelium, deposits of mineral crystals and of intratubular proteins. Interstitial inflammation is another common reaction. (Browning, 1965; Bruner, 1984; Busey and Cockrell, 1984; Clayton and Clayton, 1982; Cornish, 1980; Halder et al, 1984).

It is possible experimentally also to induce reactions in the glomerular structures by exposing to some HC. (Klavis and Drommer, 1970; Zimmerman and Norbach, 1980). These results are of great interest as, which the following text will show, it is possible that inhalation of even moderate to small amounts of HC may cause similar effects in humans.

Various types of renal neoplasms are seen after exposure to some solvents. Their relevance to discussions of human renal cancers is unclear. (Bruner, 1984; Kitchen, 1984; MacNaughton and Uddin, 1984).

Although animal studies offer advantages in terms of standardisation, substantial problems remain when comparing studies and transferring results to humans: Type and intensity of exposure as well as diagnostic criteria vary and are often irrelevant to human conditions. Furthermore, the possibility of species-specific as well as sex-specific reactions cannot be excluded. (Ballantyne, 1985; MacNaughton and Uddin, 1984).

HUMAN CASE REPORTS

More than 50 cases have been published describing acute and/or chronic renal disease following HC exposure. The most important case reports, discussing this relation have been reviewed by Churchill et al (1983). The exposure has usually been heavy as a result of accidents, suicide attempts or sniffing. It has often ended up in death from renal failure or, in later years, in dialysis treatment. Most of the main groups of HC have been represented.

The injuries are usually localised in the tubules, sometimes together with a toxic injury to the hepatic parenchyma (Cornish, 1980). Another reaction takes place in the glomerular part of the nephron, sometimes together with similar changes in the lungs. The result is a dramatic clinical picture named Goodpasture's syndrome. (Beirne and Brennan, 1972; Ehrenreich, 1977). These probably immunologic reactions in glomerular and sometimes also in similar pulmonary structures are of great interest, in the same way as the glomerular reactions of the animal experiments. They have been taken as indicators of an association between HC exposure and a specific renal disease, the glomerulonephritis (GN).

The GN is a multiphasic clinical and histopathological entity in which immunologic mechanisms probably play a prominent role. It is one of the most important renal diseases and its ethiology is to a great extent unknown. Therefore, much work has been done during the last 5-10 years in analysing this possible association between HC exposure and GN, especially as an effect of even discrete exposure cannot be excluded. The possibilities of the primary injury being situated in the tubules is also discussed. (Cameron, 1979; Ravnskov, 1985).

Not even multiple case reports can prove a relationship between for instance environmental factors and disease. The exposures, when known, vary from one case to another as do the diagnostic and therapeutic criteria and facilities. Furthermore, the influence of chance on an apparent association cannot be estimated. Case reports play however an important role in initiating other kind of studies, more suitable for estimation of causal relationships. A review of such studies will be presented.

CASE-REFERENT STUDIES

All except one of the published case-referent studies have found an excess of HC exposure in subjects with GN compared with subjects with other diseases:

Zimmerman et al (1975) presented 63 cases of renal failure. Those with renal failure because of GN had been more exposed to HC than the others with renal failure. 28 other patients in the hospital, without renal disease, had less HC exposure than those with glomerular disease.

Lagrue et al (1977) compared the HC exposure in 108 biopsy proven GN (except for 28 cases) and 54 hospitalized patients with hypertension of nephrolithiasis. There was a significant excess of HC exposure in those with GN.

Ravnskov et al (1979) made a thorough survey of exposure conditions in 50 cases with none-systemic, biopsy proven GN, 50 appendicites and 50 subjects with well defined none-glomerular renal disease. Fifty percent of those with GN had more than "slight" HC exposure, while the corresponding figure for the controls was 20 %.

Finn et al (1980) interviewed 37 cases with renal failure from GN (14 biopsy proven), 52 cases with none-glomerular renal failure and 53 male cases with none-renal disease. There was an excess of subjects with "significant" HC exposure in the GN group, compared with the other two groups.

van der Laan (1980) used the same exposure estimates as Ravnskov et al. He compared the solvent exposure in a group of 50 biopsy proven GN with 50 subjects with well defined none-renal disease. He could not find any difference in HC exposure between the two groups.

Bell et al (1985) found significantly higher organic

solvent exposure in 50 biopsy proven GN compared with 100 subjects with well defined none-renal disease. Approximately the same exposure estimates were used as by van der Laan and by Ravnskov et al.

Several objections can be raised to these studies: Considering that GN is more frequent in men, there is an unfortunate difference in sex distribution between the cases and the controls in the study by Zimmerman et al.

The paper by Ravnskov et al (1979) has recently been declared not to "meet the usual standards required for publication in medical journals". The extent of the flaws is not presented, the effect on the results can therefore not be estimated (Bengtsson, 1985).

The description of the control groups is not sufficient in the works by Zimmerman et al and by Finn et al. The possibility of a difference in some respects, for instance employability, between cases and controls, can therefore not be excluded. Such a difference might bring about various exposure opportunities. This objection may to some extent be valid also for some of the other control groups.

The control group of van der Laan's study includes diseases which often are associated with immunologic ethiology. The possibility of van der Laan showing a more general association between HC exposure and immunologic disease can therefore not be excluded.

Only 3 of the studies (Ravnskov et al, Bell et al, van der Laan) have the diagnosis of all their cases biopsy proven, the only reliable diagnostic method. They are also the only ones with well described, standardised and comparable exposure criteria: van der Laan has used the same ones as Ravnskov et al while Bell et al seem to have modified them to some extent. The cases have in no study been selected in accordance with any other study.

The effect of recall bias, i.e. a tendency of some subjects or groups of subjects to recall for instance HC exposure better than others, can probably not be excluded from any of the studies and presumably has none of the studies been performed with totally blind interviews.

It is obvious that these studies do not prove an association between varying degrees of chronic HC exposure and renal, glomerular, disease. However, design and results of some of them no doubt indicate the possibility of such an association.

COHORT STUDIES

A number of studies have also been published in which various renal function parameters were measured in organic solvent exposed subjects and in non-exposed. The results are summarised in table 1.

TABLE 1. Results of cohort studies (+ = difference between exposed and non-exposed).

Studies	1	2	3	4
Clearance	-			
Total prot		+		
Almbumin	+	-	+	+
Beta-2-m-globul	-			-
Retinol bind.prot.				-
Lysozyme		+		
Beta-glucuronidase		+		
Concentrat.capac.	-			
Cells	+			

Studies: 1=Askergren 1982 and 1984; 2=Franchini et al 1983; 3=Brochart et al 1984; 4=Lauwerys 1985.

Askergren (1982) compared 134 subjects exposed to moderate amounts of styrene, toluene and a mixture of mainly aromatics, with 48 non-exposed subjects. Five different renal fuction parameters were studied. There was no difference between the exposed and non-exposed in glomerular filtration rate, renal concentration capacity or in tubular reabsorbtion capacity of beta-2-microglobulin. There was a slight but significant increase of erythrocyte and white/tubular epithelial cell excretion as well as of albumin excretion. The studies of albumin and cell excretion have been extended to comprise 316 subjects mainly exposed to slight to moderate amounts of aromatic HC and 96 none exposed. The results are in agreement with the previous ones (Askergren, 1984).

Franchini et al (1983) have studied a large group of subjects (n=438) with mainly aliphatic and alicyclic HC exposures. They compared the urinary excretion of total protein, albumin, lysozyme, and beta-glucuronidase activity of the exposed with those of 161 non-exposed.
A summary of their results shows an increase of urinary protein and enzyme excretion in some of the exposure groups but not of albumin excretion. Their studies hence indicate rather a slight tubular than glomerular dysfunction.

Brochard et al (1984) have presented a study of proteinuria in 20 000 workers. The prevalence of proteinuria was significantly higher in those with a HC exposure than in non-exposed.

Lauwerys et al (1985) studied 180 workers moderately exposed to various HC including n-hexane, white spirit, styrene, tetrachloroethylene, methyl ethyl ketone and toluene. There was no difference, when comparing with 170 non-exposed controls, in urinary excretion of beta-2-microglobulin, retinol-binding protein or albumin except for a slight but significant shift of the cumulative frequency distribution of the urinary albumin concentration towards higher values in the styrene group. The result suggests a slight glomerular dysfunction.

A few methodologic comments should be made: Modern sensitive techniques for measuring proteins and enzymes have been used in all but one (Brochard et al, 1984). Such strips, being calibrated for clinical use, are not sensitive enough for studies of minor, praeclinical changes in urinary protein concentration. A study showing no proteinuria, measured with test strips, therefore is non-informative, while Brochard et al. may be pointing out the top of an iceberg.

Actually, there is little information on various renal function parameters' suitability for screening early dysfunction. There is also a limited knowledge concerning various analytical methods' ability to register such small quantities and variations:

Estimation of glomerular filtration rate (GFR) by means

of modern isotope technique failed to establish any change in GFR, although there was a small but significant increase in albumin excretion (Askergren, 1982).

The conventional ways of measuring tubular function do not seem useful: Estimation of renal concentrating capacity takes 36 hours of fluid deprivation to reach a maximum effect, which should be necessary to establish a minor concentrating dysfunction (de Wardener, 1956). Such a procedure is unpractical in population studies, and modern antidiuretic hormones do not seem to increase its usefulness (Askergren et al, 1981). No studies appear to have been performed estimating renal diluting or acid-base regulating capacity as screening tools.

The best evaluated methods for screening small renal dysfunctions are the modern immunologic techniques of measuring urinary protein concentration (Dubach and Schmidt, 1979; Migone et al, 1981; Weise, 1981). Unfortunately, the techniques vary, and comparisons between different methods are seldom done. This is probably one of the explanations for the wide range of normal values of protein excretion and also for the various outcomes of similar studies. Another is the great variability of study programs: Urine sample time varies, as does time of the day when urine is collected, the study populations are not comparable etc.

Much research has been done lately on urinary enzyme activity. There seem however still to exist some important objections to this technique as a screening tool of minor renal dysfunction: There is a spontaneous decrease in enzyme activity and there are several possible activity inhibitory factors present in the urine. Hence, considering the small amounts and differences to be measured, great care should still be taken when using this type of methods (Dubach and Schmidt 1979; Dubach and Le Hir, 1983; Grötsch and Mattenheimer, 1983).

HUMAN MORTALITY AND CANCER STUDIES

No analyses of associations between HC exposure and fatal outcome from nephritis or genito-urinary disease suggest such a relation (Morgan et al, 1981; Rushton and Alderson, 1981). Nor does there seem to exist any causal association in a US general population between renal carcinoma and gasoline exposure. Some studies, however, suggest that petroleum may be a risk for renal cell carcinoma and cancer of the renal pelvis and that oil refinery workers and petroleum distribution workers may be risk populations for such cancers (Higginson et al, 1984; McLaughlin, 1984; Raabe, 1984; Savitz and Moure, 1984).

The basic assumption for mortality studies is that the toxic effects are lethal. Such studies therefore hardly cast any light upon possible slight to moderate effects of exposure. Furthermore, the outcomes of "register studies" are always difficult to interpret: Diagnostic criteria may vary between and even within studies and the exposures are often vaguely or too generally defined. Such problems probably explain the inconsistence between the different studies and the sometimes inconclusive results.

SUMMARY

The final facts, necessary to establish a certain association between light to moderate organic solvent exposure and kidney disease, have not yet been presented. The methodologic difficulties and flaws from which most of the studies suffer contribute to this situation. On the other hand the results certainly do not exclude the possibility of such an association.

The mechanisms behind HC causing renal disease are obscure. A direct toxic effect does not seem probable considering the frequent exposure to HC in modern society and the fairly low incidence of this disease. The possibility of the solvents initiating, mediating or in some other way taking part in immunologic reactions appears more attractive.

The general impression after reviewing the presented investigations is, that the possibility of a causal association between moderate hydrocarbon exposure and renal disease must be considered substantial. The indications are strong enough to state that individuals with signs of renal dysfunction or manifest renal disease should not be exposed to organic solvents.

REFERENCES

Askergren A, Allgén LG, Bergström J (1981). Studies on kidney function in subjects exposed to organic solvents. II The effect of desmopressin in a concentration test and the effect of exposure to organic solvents on renal concentrating ability. Acta Med Scand 209:485-488.

Askergren A (1982). Organic solvents and kidney function. In Englund A, Ringen K, Mehlman MA (eds): "Occupational Health Hazards of Solvents". Princeton NJ: Princeton Scientific Publishers Inc, pp 157-172.

Askergren A (1984). Urinary protein and cell excretion in construction workers exposed to organic solvents. XXI Int Congr Occup Health, Dublin, September 1984.

Ballantyne B (1985). Evaluation of hazards from mixtures of chemicals in the occupational environment. J Occup Med 27:85-94.

Beirne GJ, Brennan JT (1972). Glomerulonephritis associated with hydrocarbon solvents. Mediated by antiglomerular basement membrane antibody. Arch Environ Health 25:365-369.

Bell GM, Gordon ACH, Lee P, Doig A, MacDonald MK, Thomson D, Anderton JL, Robson JS (1985). Proliferative glomerulonephritis and exposure to organic solvents. Nephron 40:161-165.

Bengtsson U (1985). Glomerulonephritis and organic solvents. Lancet II:566.

Brochard P, De Palmas J, Martini M, Blondet M, Lagrue G (1984). Etude de la prévalence des protéinuries dépistées chex des sujets exposés professionnellement aux solvants. XXI Int Congr Occup Health, Dublin, September 1984.

Browning E (1965). "Toxicity and metabolism of industrial solvents." Amsterdam, London, New York: Elsevier Publishing Company.

Bruner RH (1984). Pathologic findings in laboratory animals exposed to hydrocarbon fuels of military interest. In Mehlman MA, Hemstreet III CP, Thorpe JJ, Weaver NK (eds): "Renal Effects of Petroleum Hydrocarbons," Princeton, New Jersey: Princeton Scientific Publishers, Inc., pp 133-140.

Bulger RE (1979). Functional architecture of the kidney. In Churg J, Spargo BH, Mostofi FK, Abell MR (eds): "Kidney Disease: Present status," The Williams & Wilkins Company, pp 162-201.

Busey WM, Cockrell BY (1984). Non-neoplastic exposure-related renal lesions in rats following inhalation of unleaded gasoline vapors. In Mehlman MA, Hemstreet III CP, Thorpe JJ, Weaver NK (eds): "Renal Effects of Petroleum Hydrocarbons," Princeton, New Jersey: Princeton Scientific Publishers, Inc., pp 57-64.

Cameron JS (1979). The natural history of glomerulonephritis. In Black D, Jones NF (eds): "Renal Disease," Oxford, London, Edinburgh, Melbourne: Blackwell Scientific Publications, pp 329-382.

Churchill DN, Fine A, Gault MH (1983). Association between hydrocarbon exposure and glomerulonephritis. An appraisal of the evidence. Nephron 33:169-172.

Clayton GD, Clayton FE (eds) (1982). "Patty′s Industrial Hygienen and Toxicology." New York, Chichester, Brisbane, Toronto. Singapore: John Wiley and Sons.

Cornish HH (1980). Solvents and vapors. In Doull J, Klaassen CD, Amdur MO (eds): "Casarett and Doull′s Toxicology. The Basic Science of Poisons." New York: Macmillan Publishing Co Inc, 2nd ed, pp 468-496.

Dubach UC, Schmidt U (eds) (1979). "Diagnostic Significans of Enzymes and Proteins in the Urine". Proceedings of the International Symposium on the Biochemical Aspects of the Diagnostic Value of Urine. Kandersteg, Switzerland 1978. Bern, Stuttgart, Vienna: Hans Huber Publishers.

Dubach UC, Le Hir M (1983). Critical evaluation of the diagnostic use of urinary enzymes. III Int Symp Nephrol. Montecatini Terme 1983.

Ehrenreich T (1977). Renal disease from exposure to solvents. Ann Clin Lab Science 7:6-16.

Finn R, Fennerty AG, Ahmad R (1980). Hydrocarbon exposure and glomerulonephritis. Clin Nephrol 14:173-175.

Franchini I, Cavatorta A, Falzoi S, Lucertini S, Mutti A (1983). Early indicators of renal damage in workers exposed to organic solvents. Int Arch Occup Environ Health 52:1-9.

Grötsch H, Mattenheimer H (1983). Measurement of catalytic activity in body fluids other than blood. Urine. In Bergmeyer HU, Bergmeyer J, Grassl M (eds): "Methods of Enzymatic Analysis". Weinheim, Deerfield Beach Florida, Basel: Verlag Chemie, 3rd ed,. vol III, pp 42-49.

Halder CA, Warne TM, Hatoum NS (1984). Renal toxicity of gasoline and related petroleum naphthas in male rats. In Mehlman MA, Hemstreet III CP, Thorpe JJ, Weaver NK (eds): "Renal Effects of Petroleum Hydrocarbons," Princeton, New Jersey: Princeton Scientific Publishers, Inc., pp 73-88.

Higginson J, Muir C, Buffler PA (1984). The epidemiology of renal carcinoma in humans with a note on the effect of exposure to gasoline. In Mehlman MA, Hemstreet III CP, Thorpe JJ, Weaver NK (eds): "Renal Effects of Petroleum Hydrocarbons", Princeton, New Jersey: Princeton Scientific Publishers, Inc., pp 203-226.

Hook JB (1980). Toxic responses of the kidney. In Doull J, Klaassen CD, Amdur MO (eds): "Casarett and Doull´s Toxicology. The Basic Science of Poisons." New York: Macmillan Publishing Co Inc, 2nd ed, pp 232-245.

Kitchen DN (1984). Neoplastic renal effects of unleaded gasoline in Fischer 344 rats. In Mehlman MA, Hemstreet III CP, Thorpe JJ, Weaver NK (eds): "Renal Effects of Petroleum Hydrocarbons," Princeton, New Jersey: Princeton Scientific Publishers, Inc., pp 65-71.

Klavis G, Drommer W (1970). Goodpasture-Syndrom und Benzineinwirkung. Arch Toxikol 26:40-55.

van der Laan G (1980). Chronic glomerulonephritis and organic solvents: A case-control study. Int Arch Occup Environ Health 47:1-8.

Lagrue G, Kamalodine T, Guerrero J, Hirbec G, Zhepova F, Bernaudin JF (1977). Néphropathies glomérulaires primitives et inhalation de substances toxiques. J Urol Néphrol 4-5:323-329.

Lauwerys R, Bernard A, Viau C, Buchet JP (1985). Kidney disorders and hematoxicity from organic solvent exposure. Scand J Work Environ Health 11:suppl 1, 83-90.

MacNaughton MG, Uddin DE (1984). Toxicology of mixed distillate and high-energy synthetic fuels. In Mehlman MA, Hemstreet III CP, Thorpe JJ, Weaver NK (eds): "Renal Effects of Petroleum Hydrocarbons," Princeton, New Jersey: Princeton Scientific Publishers, Inc., pp 121-132.

McLaughlin JK (1984). Risk factors from a population-based case-control study of renal cancer. In Mehlman MA, Hemstreet III CP, Thorpe JJ, Weaver NK (eds): "Renal Effects of Petroleum Hydrocarbons", Princeton, New Jersey: Princeton Scientific Publishers, Inc,. pp 227-244.

Migone L, Scarpioni L, Cambi V (eds) (1981). "Urinary Proteins". Basel, München, Paris, London, New York, Sidney: S. Karger.

Morgan RW, Kaplan SD, Gaffey WR (1981). A general mortality study of production workers in the paint and coatings manufacturing industry. J Occup Med 23:13-21.

Raabe GK (1984). Kidney cancer epidemiology in petroleum related studies. In Mehlman MA, Hemstreet III CP, Thorpe JJ, Weaver NK (eds): "Renal Effects of Petroleum Hydrocarbons", Princeton, New Jersey: Princeton Scientific Publishers, Inc., pp 259-271.

Ravnskov U, Forsberg B, Skerfving S (1979). Glomerulonephritis and exposure to organic solvents: A case-control study. Acta Med Scand 205:575-579.

Ravnskov U (1985). Possible mechanisms of hydrocarbon-associated glomerulonephritis. Clin Nephrol 23:294-298.

Rushton L, Alderson MR (1981). An epidemiological survey of eight oil refineries in Britain. Br J Ind Med 38:225-234.

Ryan GB (1981). The glomerular sieve and the mechanisms of proteinuria. Aust NZ J Med 11:197-206.

Savitz DA, Moure R (1984). Cancer risk among oil refinery workers. A review of epidemiologic studies. J Occup Med 26:662-670.

de Wardener HE (1956). Vasopressin tannate in oil and the urine concentration test. Lancet 1:1037-1038.

Weise M (ed) (1981). "Experimental and Clinical Aspects of Proteinuria". Basel, München, Paris, London, New York, Sidney: S. Karger.

Zimmerman SW, Groehler K, Beirne GJ (1975). Hydrocarbon exposure and chronic glomerulonephritis. Lancet 2:199-201.

Zimmerman SW, Norbach DH (1980). Nephrotic effects of long-term carbon tetrachloride administration in rats. Archs Path 104:94-97.

Safety and Health Aspects of Organic Solvents, pages 169–177

SOLVENT EFFECTS ON REPRODUCTION: EXPERIMENTAL TOXICITY

György Ungváry

National Institute of Occupational Health,
Budapest, Hungary H-1450 POB 22

INTRODUCTION

The growth of world population has been forced unto a course as a result of which it will be almost doubled by the year 2000 as compared with the 1975 figures. In some European countries, on the other hand, the population decreases.

Increased chemicalization and the profound changes witnessed in the reproduction of the population may jointly lead to increasing damages in reproduction control and in the health of the offspring.

To prevent this we must experimentally identify these different kinds of damage and their characteristics.

POSSIBILITY OF TOXIC DAMAGE TO ENDOCRINE CONTROL

In the hypothalamo-pituitary system the secretion of luteotropin releasing factor (LRF) is controlled by norepinephrine, indolamines, and dopamine. LRF-secretion is also influenced by the peripheral 17 β-oestradiol (E_2) and progesterone(P) levels. Pituitary gonadotrophic hormones (FSH, LH) are controlled by LRF, E_2 and P, while gonadal function (ovarian cycle, secretion of E_2 and P) is controlled by pituitary hormones. However, endocrine control is more complicated than that. A series of interesting substances (plasma proteins, steroid binding proteins, different enzymes, proteoglycans, steroids, pituitary protein hormones, non-steroid ovarian factors) found in follicular fluid are known to influence and modify gonadal function. In addition to endocrine control, the ovarian vascular system and its autonomic (mainly noradrenergic) nerves also play an important role in the regulation of female gonad function. It is obvious that this extremely complex regulation is exposed to damage at many different points. Therefore, the endocrine system including gonadal function is more susceptible to toxic damage than other organ systems of the body.

EFFECTS OF SOLVENTS ON GONADAL FUNCTION IN NON-PREGNANT ANIMALS

Xylene and carbon disulphide are embryotoxic and have been put on the list of substances with suspected teratogenicity. Ethylbenzene - one of the components of xylene - causes anomalies in the uropoietic system of mice and rats (Ungváry and Tátrai,1985). There have been reports on disturbances of the menstrual cycle, infertility and increased frequency of spontaneous abortion in women exposed to xylene or carbon disulphide in the rubber, printing, rayon and other industries (WHO,1979; IRPTC,1984). We therefore, decided to examine whether ethylbenzene or carbon disulphide exposure can cause disturbances in the ovarian cycle.

In a fifth cycle following four normal cycles, oral doses of 0.25 or 0.5 g/kg and 0.5 or 1.0 g/kg of carbon disulphide and ethylbenzene, respectively, were given to CFY rats in the morning of oestrus, two dioestruses and pro-oestrus. At 3.00, 4.30 and 6.00 p.m on the day of the expected pro-oestrus the animals were bled to death by aortic puncture under Nembutal anaesthesia. From the blood samples E_2, P and LH levels were determined by RIA. Vaginal smears were checked daily also during treatment. From the sacrificed animals the uterus, ovary and liver were examined histologically.

On the day of the expected pro-oestrus, the vaginal smear of the exposed animals showed the picture usually seen in dioestrus. In the afternoon of pro-oestrus, in half to two-thirds of its thickness the wall of the horn of the uterus of control animal contained a wide endometrium with glands lined with proliferating tall columnar epithelium, subnuclear vacuoles (characteristic of pro-oestrus) were seen in some of the epithelial cells. The stroma of the endometrium was rich in collagenic fibres forming relatively loose bundles. In the afternoon of the expected pro-oestrus a wide endometrium was seen in the uterus of animals exposed to ethylbenzene and carbon disulphide. It contained more stroma with very dense collagenic bundles and fewer glands compared with the picture seen in the controls. The total epithelial surface of the lumen was much smaller. This, however, does not correspond to corpus luteum cysticum. In the ovary, cytolysis due to solvent exposure was seen.

At 3.00 p.m. in the afternoon of the expected pro-oestrus the LH and E_2 levels significantly decreased in all the groups of the exposed animals; it is surprising that the high dose of carbon disulphide significantly increased the P level as compared with the controls (Fig.1). From 15.00 to 18.00 hours the control value of E_2 decreased to 1/8 of the initial level. At 15.00 hours the high doses of solvents decreased the hormone level to 25%, the low doses to 50%. At 16.30 the level in the animals exposed to high doses was 50% of the level in the controls, but it had not changed as compared with the value in the same group at 15.00 hours. At 18.00 hours levels in the exposed groups decreased to half of the earlier values, but did not differ from the controls. From 15.00 hours to 18.00 hours the control values of P increased with time. At 15.00

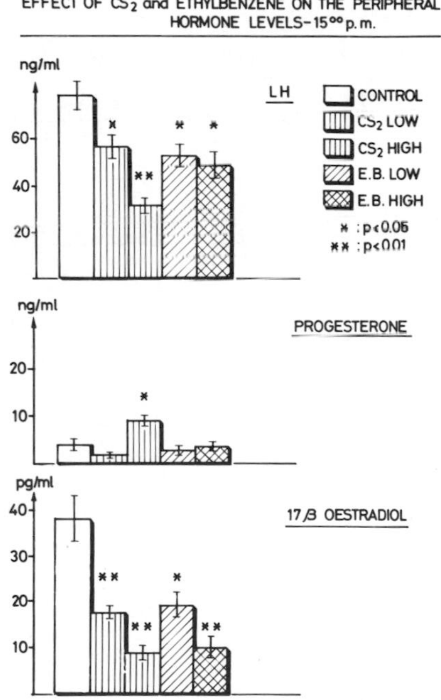

Figure 1. Effect of CS_2 and Ethylbenzene on the peripheral hormone levels - 15.00 p.m.

hours the level in the group treated with a high dose of carbon disulphide was higher than the control level. At the two other times of measurement the value was higher in both high-dose groups.

Similar hormonal alterations were elicited by exposure to carbon tetrachloride, chloroform, benzene, toluene, and xylene isomers as well.

The conclusions to be drawn from the findings are as follows:

1. Solvents block the ovarian cycle. As proved by the vaginal smears and the structure of the uterine wall, this blocking occurs during dioestrus. The absence of the change in pro-oestrus of the oestrogen-dependent vaginal smear and structure of uterine wall is understandable from the levels of E_2 which were consistently lower than the controls throughout the afternoon of the oestrus, and probably ovulation does not take place.

2. The results fail to furnish any explanation of the unexpected changes of P level. These might be due to proliferation of granulosa cells or the increased function of persisting corpora lutea. Neither of these does, however, explain the manyfold increase in the P level over the controls coinciding in time with the expected LH peak, but independently of the LH level.

WHAT IS THE EFFECT OF SOLVENTS ON THE SECRETION AND LEVEL OF SEX STEROIDS IN PERIPHERAL BLOOD OF PREGNANT ANIMALS?

In one of our experiments pregnant rats were exposed to para-xylene on the 10th, and 9th and 10th days of gestation (Ungváry et al.,1981b). Exposure to para-xylene decreased the weight of the fetuses, and the P and E_2 levels of peripheral blood, but it did not influence the uterine and ovarian venous outflow and the ovarian P and E_2 secretion rate. It is concluded that p-xylene, by inducing the hepatic MFO system (Ungváry et al.,1981a), facilitates the biotransformation of P and E_2, which are metabolized by the same system. The decrease in the sex hormone level of peripheral blood is supposed to play a role in the embryotoxicity of p-xylene.

WHY SHOULD WE INVESTIGATE THE EMBRYOTOXIC AND TERATOGENIC EFFECT OF SOLVENTS?

Solvents are of concern regarding possible embryotoxicity not only because of their widespread industrial and household uses, but also because of their generally high rates of absorption via skin, lungs, and digestive tract. High absorption is particularly characteristic of fat solvents owing to their ready passage through cellular and other biological membranes, a feature which also undoubtedly enhances passage across the placenta. Therefore, solvents are chemicals suspected to have embryotoxic and teratogenic effects.

A review of the literature by the U.S. F.D.A. (1980-cit.: Barlow and Sullivan,1982) looked at data of animal experiments on 38 chemicals for which there are reports of birth defects in humans. All, except one, had a positive study in at least one animal species, with 80% positive in more than one species. Positive responses were observed in 85% in mice, 80% in rats, 60% in rabbits. On the other hand, of 165 chemicals not known to be teratogens in humans; only 28% were negative in all species, and 50% negative in multiple species; 41% of the compounds were positive in more than one species. From practical point of view it seems to be important that about 70% of compounds had no positive effects in rabbits. No positive effects were found for these chemicals 50% of the time in rats and 35% for mice. Considering the data of FDA (1980) and our experience (Ungváry,1983), we have categorized 26 solvents, establishing the following groups based on the relevant literature:

- solvents increasing the malformation rate in two rodent species can be regarded as teratogenic with limited evidence
- solvents increasing the malformation rate in two species including rabbits can be regarded as teratogenic with sufficient evidence
- solvents not increasing the malformation rate in at least two rodents can be regarded as non-teratogenic with limited evidence
- solvents not increasing the malformation rate in three species including mice and rabbits can be regarded as non-teratogenic with sufficient evidence

Teratogenicity was established if

- the solvents and their metabolites crossed the placenta

and could be detected in fetal blood and/or amnionic fluid
the solvents increased malformation rate without sig-

TABLE 1. Teratogenic effect of 26 solvents*

Teratogens in animals		
Sufficient evidence	Limited evidence	Methodological problems
	aromatol chloroform (2B) ethylbenzene methyl-ethyl-ketone	carbon disulphide FC 22
Non-teratogens in animals		
Sufficient evidence	Limited evidence	Evaluation is not possible
benzene (1) epichlorohydrin (2B) toluene ortho-xylene meta-xylene para-xylene	carbon tetrachloride (2B) 1,1-dichloroethane dichloromethane (3) ethylene glycol monoethyl ether methyl chloroform tetrachloroethylene (3) trichloroethylene (3) xylene	aniline (3) dichlorobenzene (3) dioxane (2B) ethylene dichloride fluorocarbons (others) methyl n-butyl-ketone

*: references are not listed here
(): summary evaluation of carcinogenic risk to humans by IARC (1983)

Table 1 demonstrates the teratogenic potential of 26 organic solvents. It is conspicuous, that none of the solvents proved to be teratogenic with sufficient evidence; only 4 solvents had teratogenic effects with limited evidence. At the same time, we have sufficient or limited evidence on the absence of teratogenicity in 6 and 8 chemicals, respectively. We conclude that only a minority of the organic solvents have teratogenic activity in mammals. It is interesting that 5 out of 26 solvents have carcinogenic effect in humans and/or animals (IARC, 1983).

SIGNIFICANCE OF TOXICOKINETIC INTERACTION IN THE SUPPOSED TERATOGENIC EFFECT OF THE SOLVENTS

Knowing the metabolic pathways of solvents and of some teratogenic chemicals a potentiating interaction can be supposed between the two kinds of compounds resulting in an increase in malformation rate caused by the teratogens.

Rats were exposed to benzene, toluene, or xylene in combination with acetylsalicylic acid (ASA). All the three chemicals were found to potentiate the toxic effect of ASA. Another notable effect of the solvents was a significant increase in the frequency of malformations caused by ASA. The potentiating effect has been attributed to the crossing of the metabolic pathways of ASA and of the solvents at certain points. Benzene, toluene, and xylene, broken down to water-soluble compounds by the hepatic microsomal monooxigenase system, utilize glycine, glucuronic acid, and/or sulphuric acid during conjugation. ASA is metabolized into salicylic acid, most of which is conjugated with glycine and is excreted as hydroxy-hippuric acid. The rest is excreted in some other form (e.g. conjugated with glucuronic acid). The water--soluble compounds produced in the course of exposure to solvents (e.g. benzoic acid) deplete the glycine pool, decreasing, at the same time, the glucuronic acid level and thereby slowing down the excretion rate of ASA, while increasing the persistence and plasma concentration of salicylic acid. Our gas-chromatographic studies have shown that pretreatment with solvents increases the maternal and fetal plasma salicylic acid concentration in rats. Since the teratogenicity of ASA is due to its effect on the free salicylic acid level, the increase in the frequency of minor and major anomalies in our experiment can be attributed to the potentiating interaction between the solvents and ASA (Ungváry,1985).

Metabolic interactions may thus enhance the effect of the known teratogens. On the other hand, this interaction may remain undetected so that the increased incidence of malformations might be wrongly attributed to the agents in agricultural and industrial use. The discrepancies in the results of epidemiological surveys could be due to variations in such combined exposures at the workplace.

EVALUATION OF ANIMAL EXPERIMENTS

At present, the results of animal experiments are hardly suitable for confirming or excluding the human teratogenicity of a chemical. Therefore, the level of exposure at which the substance is still hazardous for humans cannot be exactly estimated.

WHY DO WE NEVERTHELESS RELY ON ANIMAL EXPERIMENTS IN THE FIRST PLACE?

1. Ethical considerations. It is not probable that the testing of chemicals on pregnant women will ever become acceptable.

2. The first toxicological characterization of an unknown toxic agent cannot rely on epidemiological investigations, this sholud be based on animal experiments. The results of epidemiological studies and animal experiments should be complementary, they cannot replace each other.

3. The teratogenicity of four groups of drugs which have been proved teratogenic in humans (sex steroids, thalidomide, thyreostatics, purin- or folic acid antagonist) and of the environmental chemical the teratogenicity of which has also been proved in humans (methyl mercury) has been established in animal experiments as well.

4. The aforementioned review of the literature by the U.S. F.D.A.(1980) proves that the predictive power of animal studies for positive teratogens in humans is fairly correct.

Thus the forecasts of animal experiments - if the results are positive - signalize a very well-founded suspicion of the teratogenic potential of any chemical. The predictive value of the negative results of animal experiments in both mice and rabbits seems to be important, too. A significant improvement of the postnatal test system might be expected from the introduction of methods suitable for specifically detecting the function of maturing tissues (nerve tracts and centres) and organs, and from a better exploration of the etiology and pathomechanism of teratogenesis. This is important not only for the prevention of teratogenesis, but also for the safe judgement of the harmlessness of chemicals and drugs. As a results, treatment of pregnant women will become more effective, and the number of unnecessary induced abortions or forced early retirement from work - because of mistaken judgement of the teratogenic potential of drugs or industrial or agricultural chemicals - will be reduced.

REFERENCES

Barlow SM, Sullivan FM (1982). Reproductive Hazards of Industrial Chemicals. Academic Press. London. New York. Paris. San Diego. San Francisco. Sao Paulo. Sydney. Tokyo. Toronto. pp. 9-22.

IARC (1983). World Health Organization International Agency for Research on Cancer Annual Report 1982. IARC. Lyon.

IRPTC (1984). Scientific Reviews of Soviet Literature on Toxicity and Hazards of Chemicals. N 52 Xylene. Ed.: N. F. Izmerov. Centre of International Projects, GKNT. Moscow.

Ungváry Gy (1983). Study of offspring damaging effect of chemicals (in Hungarian). Munkavédelem 29 (10-12): 207-216.

Ungváry Gy (1985). The possible contribution of industrial chemicals (organic solvents) to the incidence of congenital defects caused by teratogenic drugs and consumer goods - an experimental study. In M. Marois (ed): Progress in Clin. Biol. Res. A. R. Liss, inc., New York, Vol. 163B. pp. 295-300.

Ungváry Gy, Tátrai E (1985). On the embryotoxic effects of benzene and its alkyl derivatives in mice, rats and rabbits. Arch. Toxicol., Suppl. 8: 421-426.

Ungváry Gy, Szeberényi Sz, Tátrai E (1981a). The effect of benzene and its methyl derivatives on the MFO system. In I. Gut, M. Cikrt, G. L. Plaa (eds): Industrial and Environmental Xenobiotics. Springer Verlag. Berlin. Heidelberg. New York. pp. 285-292.

Ungváry Gy, Varga B, Horváth E, Tátrai E, Folly G (1981b). Study on the role of maternal sex steroid production and metabolism in the embryotoxicity of para-xylene. Toxicology 19: 263-268.

U. S. Food and Drug Administration (1980). Federal Register 45 (205): 69823-69824. cit. S. M. Barlow, F. M. Sullivan (1982).

WHO (1979): Environmental Health Criteria 10. Carbon disulphide. World Health Organization. Geneva.

Safety and Health Aspects of Organic Solvents, pages 179–185

SOLVENT EXPOSURE AND BIRTH DEFECTS: AN EPIDEMIOLOGIC SURVEY

Peter C. Holmberg[1)], Kari Kurppa[1)], Riitta Riala[2)], Kaarina Rantala[2)], Eeva Kuosma[1)]

[1)]Department of Epidemiology and Biostatistics, Institute of Occupational Health, SF-00290 Helsinki, Finland
[2)]Uusimaa Regional Institute of Occupational Health, SF-00290 Helsinki, Finland

INTRODUCTION

An explorative epidemiologic study on chemical and physical exposures during pregnancy of mothers of children born with selected congenital defects and of their paired referents was started in Finland in 1976. Such an investigation was thought worthwhile because noteworthy exposures can occur particularly at work sites, yet their contribution to the occurrence of congenital malformations had remained largely unexplored.

The data collection first considered the group of central nervous system (CNS) defects. The analysis of the first two years of data collecting indicated that exposure to organic solvents during the first trimester of pregnancy had occurred more often among the case mothers than among their referents (Holmberg, 1979). There was little basis a priori to expect such a result. The possibility of solvent exposure being associated with the occurrence of malformations had by then been investigated only to a limited extent. In a case series, the syndrome of caudal regression had been linked with exposure to organic solvents during pregnancy (Kucera, 1968). Controversial results regarding the risk of birth defects of children of female operating room personnel exposed to anesthetic gases have been reported, but the association has remained unconvincing; work at certain types of laboratories with mixed chemical exposure, including solvents, has also been connected with a malformation risk (for a recent review see Hemminki and Vineis, 1985).

Other groups of congenital defects (i.e., oral clefts, selected defects of the skeletal system, and selected defects of the cardiovascular system) were subsequently included in our study as well. The screening for environmental exposures during pregnancy continued until 31 December 1982 (Kurppa et al., 1983). Exposure to organic solvents appeared to have occurred more often also among the mothers of children with oral clefts when compared to the referent mothers (Holmberg et al., 1982).

In the early phases of the investigation, information on solvent exposures was scarce for many occupational activities. More exact information on Finnish exposure conditions has been furnished in the course of the study, e.g., for construction and maintenance painting (Riala et al., 1984). It has turned out that exposure to solvents can well exceed recommended threshold limit values for short term exposures (American Conference of Governmental Industrial Hygienists 1981) for instance when painting kitchen floors, walls and cabinets at homes.

The present paper reports results from a scrutiny of the total material of 1475 case-referent pairs as to the exposure to organic solvents during the first trimester of pregnancy, either at work or at home.

METHODS

Primary information about case and referent mothers was obtained through the Finnish Register of Congenital Malformations. The routines of the Register have been described in previous papers, and recently more extensively by L. Saxén affiliated to the Register since its foundation in 1963 (Saxén, 1983).

The only information on occupational factors obtained through the Register routines was confined to parental occupations and whether the mother had remained at work during pregnancy or stayed at home. Therefore, a questionnaire to illustrate specific occupational and leisure time exposures was designed (Holmberg and Nurminen, 1980). This special questionnaire with open and proforma questions was first tested for comprehensibility. It was then used throughout the present study. Data gathering was undertaken by two trained interviewers who personally interviewed all the case mothers and their referents. Most of the interviews were performed during each mother's first postnatal visit to her local Maternity Health Care Center, and a few at the home of the mother.

Information on solvent exposures was derived from: an open question; "Please describe your ordinary work day and all the different work phases" and proforma questions; "Did you have to do with solvents during the pregnancy?", "at work/ at other occasions, which solvents?".

The entire study material was first analysed as to qualitative and quantitative exposure to organic solvents blindly, i.e., without knowing the case-referent status of the mothers, by two industrial hygienists. In some instances the hygienists requested further information on exposure through personal contacts with the employers and through visits to the places of work where the exposure had occurred or through contacts to the mothers concerning domestic exposures.

After this primary work-up, the hygienists together with two experts in occupational medicine, classified the material into final exposure categories, still unaware of the case-referent status.

In the estimation of individual solvent exposures the following groups were considered: aromatic, halogenated and aliphatic hydrocarbons, lacquer petrols (mixtures of aliphatic and aromatic hydrocarbons, typical boiling range 145-200°C, containing 85 % aliphatic compounds and 15 % aromatic compounds mainly aromatic trimethyl benzene), alcohols, aldehydes, ketones, esters, and ethers. Further, mixed solvent exposures difficult to classify, e.g. selected laboratory exposures, were included.

The cut-off point for substantial exposure is inherently judgmental in explorative studies where little is known a priori about possibly hazardous exposure levels. In the quantitative classification of exposure, categories 0 to 4 were used: 0 = no exposure, 1 = "slight" exposure, concentrations less than 1/3 TLV (threshold limit values) for chemical substances in workroom air (American Conference of

Governmental Industrial Hygienists 1981), 2 = "noteworthy" exposure, when continuous concentrations about 1/3 TLV, when discontinuous peaks higher than TLV, 3 = "considerable" exposure, when continuous concentrations higher than 2/3 TLV but lower than TLV, frequent peaks higher than TLV, 4 = "heavy" exposure, long term exposure to about TLV or more. Finally, when qualitative and quantitative exposure categories had been decided the case-referent codes were opened.

RESULTS AND REMARKS

The distribution among the total study population of the 1475 case-referent pairs as to different malformation groups is given in Table 1.

TABLE 1. Study population

	No of pairs	Time of data collecting
CNS defects	365	June 1976 - December 1982
Oral clefts	581	December 1977 - December 1982
Musculoskeletal defects	360	December 1979 - December 1982
Cardiovascular defects	169	January 1980 - December 1981
Total	1 475	

The following examples may illustrate different classification categories. Quantitative classification category 1: "Dental assistants cleaning instruments with ethanol or other alcohols", "Nurses using ethanol, isopropyl alcohol, or diethyl ether as cleaning agents", "Home decorators; painting furniture with an alkyd resin paint containing 30-50 % lacquer petrol". Quantitative classification categories $\geq$ 2: "Kitchen renovation at home; painting cupboards (outside and inside) and floors using alkyd paint with 30-50 % lacquer petrol - work lasted a few days".

"Research chemist working daily with benzene, pyridine and other solvents", "Spray painter in a factory manufacturing garden furniture (main solvent xylene)", "Reinforced plastic manufacture in a boat building industry (main solvent styrene)".

Table 2 shows that exposure category 4 ("heavy" exposure) did not appear among any of the mothers in the present study. However, the case-referent distributions according to the other exposure categories indicate that exposure to organic solvents during the first trimester of pregnancy had occurred somewhat more often among the case mothers than among the referents.

TABLE 2. First trimester exposure to organic solvents

Exposure category	CNS (N=365)	Clefts (N=581)	Skeletal (N=360)	Cardio-vascular (N=169)	Total (N=1475)
Slight	24/16	29/23	17/16	9/9	79/64
Noteworthy	15/7	10/5	6/8	4/3	35/23
Considerable	2/3	5/3	2/1	2/0	11/7
Heavy	0/0	0/0	0/0	0/0	0/0

The number of "noteworthily" and "considerably" exposed mothers was 46 among the cases and 30 among the referents. Of these 24 case mothers and 14 referent mothers had been exposed when at work, while 22 case mothers and 16 referent mothers had been exposed to organic solvents during various domestic acitivities.

Lacquer petrol, toluene and xylene were the most common solvents encountered (Table 3). Among the mothers exposed outside work and classified into exposure categories $\geq$ 2, lacquer petrol was the most common exposure, and renovation and home painting the most frequent type of activity (Table 4).

TABLE 3. Mothers exposed at work

Type of work	Cases/referents	Solvents
Laboratory work	3/4	Benzene, chloroform, dichloromethane, xylene, petroleum ether, alcohols
Spray-painting	2/2	Lacquer petrol, toluene, xylene
Rubber products manufacturing	3/0	Lacquer petrol, toluene, xylene
Reinforced plastics industy	3/3	Styrene, acetone
Printing industry	6/2	Lacquer petrol, toluene, alcohols
Metal products manufacturing	4/2	Trichloroethylene, lacquer petrol
Diverse jobs/tasks	3/1	Lacquer petrol

TABLE 4. Mothers exposed outside work

Type of activity	Cases/referents	Solvents
Renovation and home painting	17/13	Lacquer petrol
Boat laminating	4/2	Styrene, acetone
Accompanying husband at home-workshop	1/1	Toluene, lacquer petrol

Unadjusted relative risk estimate for maternal first trimester solvent exposure (exposure categories $\geq$ 2) was 1.6 with 95 % confidence interval of 1.0-2.5 for all malformations pooled. Conditional logistic regression analysis yielded an identical relative risk estimate of 1.6 (95 % confidence interval 1.0–2.5) when adjusted simultaneously for potential confounding by mother's age, smoking habits, and alcohol consumption during pregnancy.

To sum up, our explorative study has afforded limited evidence to suggest a possibility for organic solvents being causally related to human teratogenesis. Such a possibility needs to be further evaluated in other sets of mothers exposed to solvents during pregnancy. Thus far, data on effects of solvent exposure during pregnancy is too sparse to allow for a well-grounded scientific inference.

REFERENCES

Hemminki K, Vineis P (1985). Extrapolation of the evidence on teratogenicity of chemicals between humans and experimental animals: Chemicals other than drugs. Teratogenesis Carcinogenesis Mutagenesis 5: 251–318.

Holmberg PC (1979). Central-nervous-system defects in children born to mothers exposed to organic solvents during pregnancy. Lancet 2: 177–179.

Holmberg PC, Hernberg S, Kurppa K, Rantala K, Riala R (1982). Oral clefts and organic solvent exposure during pregnancy. Int Arch Environ Health 50: 371–376.

Holmberg PC, Nurminen M (1980). Congenital defects of the central nervous system and occupational factors during pregnancy. A case-referent study. Am J Ind Med 1: 167–176.

Kucera J (1968). Exposure to fat solvents: a possible cause of sacral agenesis in man. J Pediatr 72: 857–859.

Kurppa K, Holmberg PC, Hernberg S, Rantala K, Riala R, Nurminen T (1983). Screening for occupational exposures and congenital malformations. Preliminary results from a nationwide case-referent study. Scand J Work Environ Health 9: 89–93.

Riala R, Kalliokoski P, Pyy L, Wickström G (1984). Solvent exposure in construction and maintenance painting. Scand J Work Environ Health 10: 263–266.

Saxén L (1983). Twenty years of study of the etiology of congenital malformations in Finland. In Kalter H (ed): "Issues and Reviews in Teratology," vol 1, New York: Plenum Press Corporation, pp 73–110.

Safety and Health Aspects of Organic Solvents, pages 187–202

CANCER HAZARD FROM EXPOSURE TO SOLVENTS

Paul Grasso

Robens Institute
University of Surrey
Guildford Surrey GU2 5XH

Cancer accounts for a third to a quarter of all deaths in our time (Am Cancer Soc) and hence it is important to take adequate steps to prevent the disease as well as to direct efforts to its early diagnosis and cure. Prevention can only be applied if the environmental causes of cancer can be identified and a considerable effort is being directed at the moment towards this end. There are two principal methods available for assessing the carcinogenic hazard to man, firstly the identification of a chemical as a carcinogen and secondly the monitoring of human exposure to it.

The methods available for the detection of chemical carcinogens are limited to two or possibly three approaches. The first approach is the epidemiological one, the second is animal experimentation and the third method, which is likely to acquire more importance in the future, is the in vitro group of tests known generically as short-term tests for carcinogenicity and mutagenicity.

In this talk we shall touch briefly on some of the epidemiological studies employed to discover the carcinogenic activity of solvents but attention will be mainly paid to animal experiments and particulrly to the pitfalls that may be encountered in assessing the carcinogenic risk from the results of such studies.

Let us first of all have a look at the reasons for using animals in the detection of carcinogens. The animals employed are mainly of the mammalian species -

non-mammalian species (eg, fish) have occasionally been employed but experience with them is very limited. For logistic and cost reasons, mice and rats are the species of choice but where the situation demands it, higher species, including primates, are used as well.

Experience shows that all mammals develop cancer towards the end of their life-span. In this respect they resemble man. The tumours that develop in animals may run a benign or malignant course as in man and, as in man, they may arise in any cell type. Histologically, tumours resemble also the human ones in architecture although but the histological appearance of the tumour is not necessarily associated with the same type of biological behaviour in animals and in man (Turusov, 1976, 1979). Furthermore, it has been repeatedly shown that substances which are carcinogenic to man are also carcinogenic in one or more species of laboratory animals. In Table 1, I have summarised some well known causes of human cancer. All of these chemicals and products (with one or two exceptions) have been tested in rodents or in dogs and gave a positive result when tested in animals.

This experience has served to indicate that animals are useful models for detecting substances that are likely to present a carcinogenic hazard for man and over the years many chemicals (including some solvents) have been tested by one or more route and found to be carcinogenic. Unfortunately, however, several pitfalls have been encountered in extrapolating animal data to the human and, in discussing the carcinogenicity of solvents, some of these pitfalls will be mentioned.

Although epidemiological studies have been conducted on a number of solvents, a reasonably firm connection between solvent exposure and cancer was only obtained in the case of benzene. It is interesting to trace the historical background which led to the acceptance of benzene as a human carcinogen since it typifies the way in which human evidence is collected and analysed.

More than half-a-century ago Delore and Borgomano

Table 1

EXAMPLES OF CHEMICALS THAT ARE CARCINOGENIC IN BOTH MAN AND EXPERIMENTAL ANIMALS

Chemical	Man	Rat	Mouse	Hamster	Dog	Other Species
4-aminobiphenyl	Bladder	Mammary Gland Intestine	Bladder	-	Bladder	Rabbit - Bladder
Asbestos	Lung, pleura Gastro intestinal tract	Lung, pleura	Lung, pleura	Lung, pleura	-	Rabbit - Lung, pleura
Benzene	Bone-marrow	-	-ve	-	-	-
Benzedine	Bladder	Liver Zymbal gland Colon	Liver	Liver	Bladder	-
BCME	Lung	-	Lung Nasal cavity Skin	-	-	-
Chromium (chromate production	Lung Nasal cavity	Lung	-	-	-	-
Haematite mining	Lung	- ve	- ve	- ve	-	Guinea-pig - negative
Mustard gas	Lung Larynx	-	Lung Mammary gland	-	-	-
2-naphthylamine	Bladder	-	Liver Lung	Bladder	Bladder	Monkey - bladder
Nickel (refining)	Nasal Cavity Lung	Lung	-	-	-	-
Soot, tars and oils	Lung Skin	-	Skin	-	-	Rabbit - skin
Vinyl Chloride	Liver Lung	Liver Lung Kidney	Liver Lung Kidney	-	-	-

Note: Dash (-) means no relevant information available
Fronm IARC Monographs on the Evaluation of the Carcinogenic Risk of Chemicals to Humans. Supplement 1, IARC Lyon 1979.

(1928) drew attention to a possible association between leukaemia and occupational exposure to high concentrations of benzene. A few years later, Emile-Weil (1932) made similar observations. With one exception, all these cases had occurred in workers occupationally exposed to high concentrations of benzene vapour, usually of the order of 150-1,000 ppm. Many "quasi-epidemiological" studies have been reported on hospital patients in whom a diagnosis of leukaemia has been correlated retrospectively with an occupational history of exposure to benzene while examination of groups of workers with known high exposures to benzene revealed a higher than usual incidence of leukaemia (Van Raalte and Grasso, 1982). Since then the levels of benzene have gradually been lowered and the current exposure seldom exceeds 5-10 ppm but well-conducted epidemiological studies carried out on workers who had been exposed to benzene before current hygienic measures were enforced have demonstrated a higher than expected incidence of leukaemia: the cases of leukaemia were found to have occurred amongst those workers whose jobs involved a very high (>1,000 ppm) exposure to benzene (Van Raalte and Grasso, 1982). Further epidemiological studies have been carried out on workers whose exposure was much less than the "heroic" concentrations encountered fifty or so years ago. As far as can be judged at the moment, there is no convincing evidence that exposure to concentrations of benzene below 100 ppm have been causally connected with leukaemia.

I think there is a point worth making in considering the carcinogenicity of benzene. It is usually assumed in these epidemiological studies that the only mode of human exposure was by inhalation and yet it is inconceivable that some skin absorption had not taken place from the exposed parts of the body such as the hands and face, so that in reality, the actual dose of benzene absorbed is probably considerable higher than the epidemiological studies suggest. Although this consideration may provide some additional degree of reassurance that benzene is not as potent a carcinogen for man as some studies would suggest, there is no room for complacency and efforts should be made to reduce exposure to the lowest practicable level.

Benzene appears to be an exception to the rule that human carcinogens are also carcinogenic to animals. Several attempts have been made in the past to reproduce in animals the bone-marrow effects seen in man. Although bone-marrow depression and aplasia have been produced readily in experimental species (Van Raalte and Grasso, 1982),it has not been possible to produce leukaemia in these species. Recently, however, Maltoni has succeeded in inducing ear duct tumours (Zymbal gland carcinomas) in Sprague-Dawley rats, by excessively high doses of benzene given orally. Although this brings benzene into line with other human carcinogens in their ability to induce tumours in animals as well as in man, it does not explain why rodents are refractory to the leukaemogenic effect of this chemical.

There is hardly any epidemiological work to speak of on the carcinogenicity of other solvents (See IARC, 1979) so that the animal carcinogenicity data are the only source of information on which to base an assessment of carcinogenic hazard to man. In fact very few of the 300 or so solvents in current use have been tested for carcinogenicity. One assumes that these few were selected either because of a history of heavy exposure and/or because of extensive industrial use.

The commonly used solvents that have been tested for carcinogenicity in rats and mice are listed in Table 2 and all of them produced a statistically significant increased incidence of tumours and by definition they are therefore, carcinogenic to experimental animals. Eight of these solvents produced liver tumours, one produced tumours in the stomach and mammary gland and one produced tumours in the urinary bladder. Only three solvents, acetone, toluene and dichloromethane have been tested with negative results (Van Duuren et al, 1978, Coombs, 1978, Theiss et al, 1977).

Most of these solvents have been in use for a considerable number of years and one suspects that exposure of workers to many of them was not much less than that to benzene and yet there is no indication that any of them have produced cancer in man. It is, therefore, essential to examine critically the results of the carcinogenicity tests in order to determine the relevance to human hazard of the type of tumours induced in rodents by these solvents.

Liver Tumours in Rodents

Tumours of the liver are induced more frequently than tumours in any other organ of the rodent by chemical carcinogens (ECETOC, 1982). A large proportion of hepatocarcinogens in rodents are non-genotoxic and include most of the chlorinated hydrocarbon carcinogens and substances such as phenobarbitone (Schulte-Hermann, 1974), BHT (Olsen et al, 1983) and selenium (Wilbur, 1980). The solvents that induced liver tumours are either non-genotoxic or weakly mutagenic. Seemingly they produced mutations only in bacteria, [with the exception of 1,2-dichloroethane which induced mutations in Drosophila as well] (See IARC, 1979). It is relevant to examine briefly some of the mechanisms by which non-genotoxic chemicals induce liver tumours.

Most, if not all, genotoxic hepatocarcinogens either produce hepatocellular necrosis or some type of sub-cellular change such as SER hypertrophy, peroxisome proliferation or increased lysosomal activity (ECETOC, 1978, Schulte-Hermann, 1974, Cohen and Grasso, 1981). Hepatocellular necrosis stimulates a reparative process which invloves active cell proliferation. Continued damage to the liver leads to a cycle of cell necrosis and regeneration. It is generally considered that cells in an active phase of division are more prone to transformation than normal cells (Coombs and Bhatt, 1978) either because they are more sensitive to damage from environmental carcinogens or else because there is a greater chance of an error in DNA replication if cells are forced to proliferate more frequently and over a much longer period than under conditions of "fair wear and tear" (ECETOC, 1982). The increased turnover rate of hepatocytes probably accounts for the development of cancer when repeated cycles of necrosis and regeneration are provoked.

The sub-cellular changes in the liver mentioned in the previous paragraph are almost invariably associated with liver enlargement, a process which involves an initial phase of hyperplasia. If the liver is maintained in an enlarged state by the continued administration of the chemical a subsequent and gradual increase in ploidy levels occurs (Schulte-Hermann, 1974) by a process of endonuclear reduplication. This process is considered to be the counterpart of hyperplasia (Brodsky and Uryvaeva, 1977).

The continued replication of DNA which endonuclear reduplication entails would appear to be as prone to lead to the development of transformed cells as the replication that takes place prior to cell division since the majority of compounds that cause liver enlargement also cause tumours (Schulte-Hermann, 1974, Cohen and Grasso, 1981). At current levels of exposure of man to solvents it is most unlikely that liver cell necrosis would occur. Furthermore, the phenomenon of liver enlargement so common in the rodent, does not appear to occur in man (Cohen and Grasso, 1981), so that the risk of cancer development in the liver of man from hepatocarcinogenic non-genotoxic solvents at current levels of exposure is negligible.

Solvents have induced other types of tumours apart from liver tumours (See Table 2). For example, 1,2-dichloroethane has induced squamous cell carcinoma in the stomach, haemangiosarcomas and uterine tumours in rats and mammary gland tumours, lymphomas and pulmonary tumours in mice, while diethylene glycol has induced tumours of the bladder.

Significance of Stomach Tumours

The stomach in rats is made up of a squamous (forestomach) and a glandular portion which are anatomically divided by a ridge of tissue known as the limiting ridge. The tumours induced by 1,2-dichloroethane were squamous cell carcinomas and therefore arose in the forestomach. The stomach of the mouse has a similar type of anatomical structure as that of the rat.

Earlier work, particularly in the mouse, has shown that this portion of the stomach is sensitive to tumour production by polycyclic aromatic hydrocarbons and other classes of carcinogens given by the oral route (Odashima, 1979). Furthermore, naturally occuring tumours are exceedingly rare in this region (Odashima, 1974) so that any tumour induction here has a much more clear relationship to the inducing agent than tumours induced in other tissues where the natural incidence is high (eg, 5% or more).

On this basis, the induction of tumours of the forestomach by 1-2,dichloroethane is clear evidence of its carcinogenicity. The significance for man of tumour

Table 2

SOLVENTS TESTED FOR CARCINOGENICITY WITH POSITIVE RESULTS

Solvent	Route of administration	Species	Site of carcinogenic action	Ref
Carbon tetra-chloride	Oral inhalation and subcutaneous	Mouse, rat and hamster	Liver	1
Chloroform	Oral	Mouse and Rat	"	2
1,4-Dioxane	"	Rat	"	3
Hexachloro-ethane	"	Mouse	"	4
1,1,2-Tri-chloroethane	"	"	"	5
1,1,2,2-Tetra-chloroethane	"	"	"	6
Trichloro-ethylene	"	"	"	7
Tetrachloro-ethylene	"	"	"	8
1,2-Dichloro-ethane	"	Rat	Forestomach Mammary glands Haemangio-sarcoma	9
"	"	Mouse	Mammary glands Liver, Lung Uterus	9
Diethylene glycol	"	Rat	Bladder	10

induction of this sort is, however, debateable. It has been shown that propionic acid (Mori, 1953) and butylated hydroxyanisol (BHA) (Ito et al, 1983) can induce tumours in this region. Propionic acid is a product of intermediary metabolism, while BHA had been tested previously for carcinogenicity mixed with the diet with negative results (Deichmann, 1955). Studies of the early changes with BHA (P Grasso's personal observation) have revealed marked hyperplasia in the squamous epithelium suggesting an increased turn-over rate of the cells in this tissue. As discussed in the previous section this increased turn-over rate could be the principal factor in cell transformation in this epithelium as in liver.

Mammary Tumours in Mice

1,2-dichloroethane induced also mammary tumours in mice. Mammary tumours in this species are now held to be caused principally by the mammary tumour virus (MTV) (originally was known as the Bittner agent) since most strains of mice are infected by it (Grasso et al, 1977). The production of tumours by this agent, however, is influenced by two other factors: genetic tendency and hormonal status. The interplay of these three factors determines the background incidence of tumours in most of the mouse strains used in conventional carcinogenicity screening tests and accounts for strains with a low incidence as well as strains with a high incidence of tumours. Administration of carcinogens, eg, polycyclic aromatic hydrocarbons, by the oral route can increase considerably the incidence of these tumours but there is no evidence that these compounds are acting solely on the genome (Grasso et al, 1977). Since oestrogenic hormones, both the synthetic and naturally occuring ones increase considerably the incidence of these tumours (Nandi and McGrath, 1973) the carcinogens could be exerting some hormone-like action on the mammary gland tissue. Thus, the induction of mammary gland tumours is an uncertain index of carcinogenicity and cannot be advanced in support of any claims for the carcinogenic activity of 1,2-dichloroethane.

Significance of Pulmonary Tumours in Mice

These were induced by 1,2-dichloroethane and there was also a marginal increase in these tumours in mice treated with dichloromethane.

The factors responsible for the induction of tumours in mice were discussed in detail in a review by Shimkin and Stoner (1975). These authors cited evidence which indicates that these tumours are predominantly benign and are genetically determined. It would thus appear that some of the lung cells in the mouse are prone to undergo neoplastic transformation. This subject is discussed in detail in IARC Mono 20 and the conclusion reached was that induction of these tumours provides only a limited evidence for carcinogenicity so that on their own, they are unsuitable for assessing hazard to man.

Carcinomas of the Urinary Bladder

These tumours were reported in the results of carcinogenicity tests of diethylene glycol (Weil et al, 1965). This chemical was at one time extensively used in the pharmaceutical industry and to a more limited extent as a solvent in the food industry. It may occur as a contaminant of polyoxyethyleneglycol, an emulsifying agent used in food and may, therefore, be ingested by a wide section of the population as well as presenting some hazard to those handling the product.

Weil et al (1965), pointed out that stones were found in all the bladders bearing tumours and demostrated that the bladder tumours did not develop in the absence of stones. Furthermore, they demonstrated that the stones produced tumours when implanted in the bladder of otherwise untreated rats.

Later experiments demonstrated that foreign bodies, such as paraffin wax or cholesterol pellets or glass beads, produced vesical carcinoma (Grasso, 1970). Attempts were made to explain these tumours on carcinogenic contaminants, but even glass beads that were carefully washed with soap and water and then rinsed in several changes of distilled water produced carcinomas. A study of the early changes produced by foreign bodies implanted in the vesical lumen revealed that a marked hyperplasia developed soon after implantation and that a strong correlation exists between hyperplasia and the later development of carcinoma (Grasso, 1976).

This correlation was established more firmly following the publication of the results of an experiment on ethyl-sulphonyl naphthalene 1-sulfonamide (ENS) by Flaks et al, (1973). This compound is a potent bladder carcinogen in mice and a number of experiments have shown that it also produces pronounced hyperplastic changes as well as calculi in this organ. The calculi are thought to result from the alkaline urine produced by the administration of ENS. When the urine was acidified by the addition of ammonium chloride to the drinking water, neither calculi or tumours of the bladder were produced by ENS administration (Flaks et al, 1973).

The sum total of these observations indicate that if the compound produces bladder calculi then no conclusion can be drawn on the carcinogenic properties of the compound.

Significance of Lymphomas in Mice

It is now acknowledged that most mouse strains that are used in carcinogenicity testing harbour a variety of lymphoma viruses. This topic has been reviewed by Lilly and Pincus (1973) who concluded that two interrelated factors are responsible, namely the presence of a virus belonging to the C-type RNA group, a gene (or genes) that govern the likelihood of infection by lymphoma viruses and the host response to the transformed cell. The administration of a chemical may in some instances tip the scales against the host by, for example, interfering with the immune defence mechanism, allowing the virus to produce a tumour.

Because genetics and tumour viruses may play a major role in the induction of lymphomas, no reliance can be placed on an increased incidence of this tumour as evidence of carcinogenicity.

CONCLUSION

Of the solvents that have been investigated for their carcinogenic potential only benzene presents any real hazard. There is ample human evidence that exposure to levels of benzene of 100 ppm and over in ambiet air carries a risk of producing leukaemia. None of the other solvents present any real hazard. As I have attempted to show, there are grounds for suspecting that the tumours observed when these solvents were tested in animals are not induced as a result of damage to the genome (with the possible exception of 1,2-dichloroethane but by some other mechanism such an increase in cell turn-over rate or an increase in nuclear ploidy, viral infection, genetic factors or hormonal status. Although there are grounds for some reassurance that these chemicals are not carcinogenic for man every effort should be made to reduce human exposure to the lowest limit practicable.

1,2-dichloroethane deserves some special attention. This compound seems to have produced a galaxy of tumours, three of which (stomach, lung and mammary gland) have been commented upon. The other two tumour types (haemangiosarcoma and uterine tumours) are uncommon in control rats. For some unknown reason the haemangiosarcomas were found in a variety of tissues rather than in one organ as, say, vinyl chloride (liver) or propylene oxide (nasal mucosa). It could be argued that these tumours arise from the connective tissue and since this is diffusely distributed thoughout the body the scatter of haemangiosarcoma is a reflecion of the anatomical distribution of connective tissue. Although this argument is a sound one from the histogenetic viewpoint it must be kept in mind that the haemangiosarcomas were not dose-related and of a relatively low incidence. Only in the males did the tumour incidence reach statistical significance. The former comment applies also to the uterine tumours so that although the cited evidence provides a formidable array of tumours the hazard to man is not as great as the numbers and variety of the tumours might suggest. Nevertheless, the experimental findings with this solvent indicate that it is likely to present a greater hazard than any of the other solvents discussed in this talk, except for benzene.

REFERENCES

American Cancer Society. Cancer facts and figures - New York. American Cancer Society 1974

Brodsky WY and Uryvaeva IV (1977). Cell ploidy: its relation to tissue growth and function. Int Rev Cytol 50: 275

Cohen AJ and Grasso P (1981). Review of the hepatic response to hypolipidaemic drugs in rodents and assessment of its toxicological significance to man. Fd Cosmet Toxicol 19: 585-605

Coombs MM and Bhatt TS (1978). Lack of initiating activity in mutagens which are not carcinogenic. Brit J Cancer 38: 148-150

Deichmann WB, Clemmer JJ, Rakoczy R and Bianchine J (1955). Toxicity of ditertiarybutylmethyl-phenol. Arch Ind Hlth 11: 193

Delore P and Borgomano C (1928). Acute leukaemia in the course of benzene intoxication: on the toxic origin of certain acute leukaemias and their relationship to serious anaemias. J Med, Lyon 9: 227-233

ECETOC (1982). Hepatocarcinogenesis in laboratory rodents: relevance for man.

Emile-Weil MP (1932) La leucemie benzolique. Bull Mem Soc Med Hop, Paris 46: 193-198

Flaks A, Hamilton JM and Clayson DB (1973). Effect of ammonium chloride on incidence of bladder tumours induced by 4-ethylsulfonylnaphthalene-1-sulphonamide. JNCI 51: 2007

Grasso P (1970). Carcinogenicity testing and permitted lists. Chemistry in Britain 6: 17-22

Grasso P (1976). Reviews of tests for carcinogenicity and their significance to man. Clinical Toxicology 9: 745-760

Grasso P, Crampton RF and Hooson J (1977). In "The mouse and carcinogenicity testing". BIBRA, Surrey

IARC (1979). Monographs on the evaluation of the carcinogenic risk of chemicals to humans. Some Halogenated Hydrocarbons 20: IARC, Lyon

Ito N, Fukushima A, Hagiwara H, Shibata M and Ogiso T (1983). Carcinogenicity of butylated hydroxyanisole in F344 rats. Cancer Res 70. 343

Lilly F and Pincus T (1973). Genetic control of murine viral leukemogenesis. Adv Cancer Res 21:1

Mori K (1953). Production of gastric lesions in the rat by the diet containing fatty acids. Gann 44: 421-426

Nandi S and McGrath CM (1973). Mammary neoplasia in mice. Adv Cancer Res 17: 353

Odashima S (1979). Tumours of the oral cavity, pharynx, oesophagus and stomach. In "Pathology of Tumours in Laboratory Animals" Vol 2 Tumours of the Mouse. Editor-in-chief VS Turusov. IARC Scientific Publ No 23, Lyon

Olsen P, Bille N and Meyer O (1983). Hepatocellular neoplasms in rats induced by butylated hydroxytoluene Acta Pharmacol et Toxicol 53: 433-434

Pathology of Tumours in Laboratory Animals. Vol 1 parts one and two - Tumours of the Rat. IARC Scientific Publ 5 (1973) and 6 (1976). Vol 2 - Tumours of the Mouse. IARC Scientific Publ 23 (1979). Editor-in-chief VS Turusov. IARC, Lyon

Schulte-Hermann R (1974). Induction of liver growth by xenobiotic compounds and other stimuli. CRC Critical Rev in Toxicology 3: 97-158

Shimkin MB and Stoner GD (1975). Lung tumours in mice: Application to carcinogenesis bioassay. Adv Cancer Res 21: 1

Theiss JC, Stoner GD, Shimkin MB and Weisburger EK (1977). Test for the carcinogenicity of organic contaminants of United States drinking waters by pulmonary tumour response in Strain A mice. Cancer Res 37: 2717-2720

Van Duuren BL, Loewengart G, Seidman I, Smith AC and Melchionne S (1978). Mouse skin carcinogenicity tests on flame retardants tris(2,3-dibromopropyl)phosphate, tetrakis(hydroxymethyl)phosphonium chloride and polyvinyl bromide. Cancer Res 38: 3236-3240

Van Raalte HGS and Grasso P (1982). Haematological, myelotoxic, clastogenic, carcinogenic and leukaemogenic effects of benzene. Regulatory Toxicology and Pharmacology 2: 153-176

Wilbur GC (1980). Toxicology of selenium - A review. Clinical Toxicology 17: 171-230

REFERENCES (Table 2)

1 NIOSH (1976). Criteria document: recommendations for a carbon tetrachloride standard. Occupational Safety Health Report 5: 1247-1253

2 National Cancer Institute (1976). Report on the carcinogenesis bioassay of chloroform, Bethesda, MD, Carcinogenesis Program, Division of Cancer Cause and Prevention

3 IARC (1976). Monographs on the evaluation of carcinogenic risk of chemicals to man. Vol II, Lyon

4 NIOSH (1978). Current Intelligence Bulletin No 20

5 National Cancer Institute (1978). Bioassay of 1,1,2-Trichlorethane for possible Carcinogenicity (Tech Report Series No 74). DHEW Publication No (NIH) 78-1324. Washington DC, US Dept of Health, Education and Welfare

6 National Cancer Institute (1978). Bioassay of 1,1,2,2-Tetrachlorethane for possible Carcinogenicity. DHEW Publication No (NIH) 78-827. Washington DC, US Dept of Health, Education and Welfare

7 National Cancer Institute (1976). Carcinogenesis Bioassay of Trichloroethylene (Tech Report Series No 2). DHEW Publication No (NIH) 76-802. Washington DC, US Department of Health, Education and Welfare.

8 National Cancer Institute (1977). Bioassay of Tetrachlorethylene for possible Carcinogenicity (Tech Report Series No 13). DHEW Publication No (NIH) 77-813. Washington DC, US Dept of Health, Education and Welfare

9 National Cancer Institute (1978). Bioassay of 1,2-Dichloroethane for possible carcinogenicity (Tech Report SEries No 55). DHEW Puublication No (NIH) 78-1361. Washington DC, US Dept of Health, Education and Welfare

10 Weil CS, Carpenter CP and Smyth HF (1965). Urinary bladder response to diethylene glycol. Calculi and tumours following repeated feeding and implants. Arch Envir Health 11: 569

Safety and Health Aspects of Organic Solvents, pages 203–224

APPLICATION OF PSYCHOMETRIC TECHNIQUES IN THE ASSESSMENT OF SOLVENT TOXICITY

Francesco Gamberale

Research Unit of Psychophysiology
Occupational Health Department
National Board of Occupational Safety and Health
Solna, Sweden

INTRODUCTION

Many hypoteses have been formulated and tested in empirical investigations concerning the nature of the effects to be expected in solvent exposure. These hypotheses cover a wide spectrum of impairments of the functional capacity of the nervous system. They range from a transitory depressant action on arousal, to serious irreversible organic brain dysfunctions.

The choice of the behavioral performance tests should depend on the hypothesis to be tested. A slight transient effect on arousal during short-term exposure to low solvent concentration could probably only be detected by means of a very sensitive and reliable vigilance task. In the case of serious organic brain damage, on the other hand, almost any behavioral performance test would probably reveal a deficit in mental capacity. In the latter case, the sensivity of the tests to detect solvent-induced performance changes would be of secondary importance More important would be to evaluate the extent and nature of the damage, which would require the use of a comprehensive and differentiated battery of behavioral tests.

The present paper presents a short review of the main results obtained from the application of behavioral performance tests in the study of solvent toxicity and is based on a paper presented at the International Conference on Organic Solvent Toxicity held in Stockholm 1984 (Gamberale, 1985). The studies reviewed have been divided into four categories depending on whether they were 1) experimental laboratory investigations, 2) quasi-experimental field studies, 3) post factum or epidemiological studies and 4) clinical studies or studies using data from clinical investigations.

Some of the methodological characteristics of the above mentioned types of investigations are indicated in Table 1. They will briefly be discussed in the paper in connection with the presentation of the results from each of the four categories of investigations.

Table 1. Types of investigations of solvent-induced effects on behavioral performance.

TYPE OF STUDY	CONDITIONS	SUBJECTS	TYPE OF EFFECT
Experimental study	Manipulation and control of exposure	One group Repeated measurements	Short-term effect on CNS
Quasi-experimental field study	Control of exposure	Two groups Repeated measurements	Short-term effect on CNS
Post factum study	No control of exposure	Two or more groups	Long-term effect on CNS
Clinical study	No control of exposure	N = 1	Long-term effect on CNS

EXPERIMENTAL LABORATORY STUDIES

A general feature of all human experimental studies in behavioral toxicology is the examination by means of behaviortal performance tests of the acute effects on the nervous system of exposure to a neurotoxic substance under conditions which are manipulated and controlled by the researcher. With an adequate experimental design and procedure it is possible to eliminate or control the potential effects of confounding factors, i.e. factors which, although extraneous to the experiment, may affect performance. If properly performed, the experiment may provide results which directly can be interpreted in terms of cause-effect relationship between exposure and performance.

Due to the fact that human experimental studies are both very expensive and difficult to conduct, they are few in number. It is therefore possible to produce a fairly representative illustration of the results obtained even in a short review such as the present one. Table 2 illustrates some of the features and results of 35 human inhalation experiments in which behavioral performance tests constituted the main criteria in the assessment of solvent toxicity. Experiments in which the evaluation of toxicity was primarily based on neurophysiological variables are not included. For each experiment, Table 2 lists the type of solvent investigated, the duration of exposure and the number of exposed subjects. Regarding the outcome of the experiments, the table reports the lowest solvent concentration at which behavioral performance impairments were observed as well as the highest concentration not resulting in manifest decrements in performance.

As can be seen in Table 2 the duration of exposure in the different experiments ranges between 1 and 8 hours. This is a considerable variation and should be taken into account when comparing the results. From an analysis of Table 2 some facts emerge. Behavioral performance was affected by exposure to relatively low concentrations of most of the solvents investigated, i.e. methylene chloride, styrene, toluene, 1,1,1-trichloroethane, trichloroethylene, white spirit and xylene.

Although the concentrations leading to manifest performance decrements are sometimes encountered in the work environment, they do appear to be above the present TLVs

Table 2. Experimental laboratory investigations of solvent-induced effects on behavioral performance.

AUTHOR	SOLVENT	NO OF SUBJECTS	DURATION OF EXPOSURE	LOWEST CONCENTRATION WITH MANIFEST PERFORMANCE DECREMENT ppm	HIGHEST CONCENTRATION WITH NO MANIFEST PERFORMANCE DECREMENT ppm
Putz-Anderson et al 1981	Methyl Chloride	56	3 hours		200
Di Vincenzo et al 1972	Methylene Chloride	11	2 hours		200
Winneke 1974	Methylene Chloride	12 and 18	3 and 4 hours	300	
Gamberale et al 1975	Methylene Chloride	14	4 x 30 min		800
Putz et al 1979	Methylene Chloride	12	4 hours	200	
Stewart et al 1968	Styrene	5	60 min	375	200
Gamberale et al 1974	Styrene	12	4 x 30 min	350	250
Stewart et al 1970	Tetrachloro-etylene	17	7 hours		100
Stewart et al 1977	Tetrachloro-ethylene	12	5.5 hours		100
	Alcohol	12		0.6 ml/kg	
	Diazepam	12		10 mg	
	Combinations	12		No interactions	
Gamberale et al 1972	Toluene	12	4 x 20 min	300	100
Stewart et al 1975	Toluene	8	7,5 hours		100
Winneke 1982	Toluene	18	3.5 hours		100
Andersen et al 1983	Toluene	16	6 hours		100
Cherry et al 1983	Toluene	8	4 hours		80
	Alcohol	8	4 hours	0.4 ml/kg	
	Combination	8	4 hours	0.4 ml and 80 ppm toluene (no interaction)	
Dick et al 1984	Toluene	30	4 hours		100
	M E K	20	4 hours		200
	Combination	20	4 hours		100 (Toluene) 50 (M E K)

Anshelm Olson et al 1984	Toluene	16	4 hours		80
	Xylene	16	4 hours		80
	Combination	16	4 hours		50 toluene + 30 xylene
Iregren et al 1984	Toluene	12	4.5 hours		80
	Alcohol	12	4.5 hours	0.6 ml/kg	
	Combination	12	4.5 hours	0.6 ml and 80 ppm toluene (no interaction)	
Stewart et al 1969	Trichloro-ethane	5	7 hours	500	
Salvini et al 1971	Trichloro-ethane	6	2 x 4 hours		450
Gamberale et al 1973	Trichloro-ethane	12	4 x 30 min	350	250
Vernon and Fergusson, 1969	Trichloro-ethylene (TCE)	8	2 hours	1000	300
Fergusson and Vernon, 1970	TCE and depressant drugs (alcohol)	8	2 hours	300 ppm TCE and 0.5 ml/kg	
Salvini et al 1971	TCE	6	8 hours	110	
Stewart et al 1974	TCE	9	8 hours		110
Nakaaki et al 1973	TCE	4	6 hours		100
Konietzko et al 1975	TCE	20	4 hours		95
Ettema et al 1975	TCE	15	2.5 hours		300
Windemuller and Ettema, 1978	TCE Alcohol	15	2.5 hours		200 ppm TCE and 0.4 ml/kg
Gamberale et al 1976	TCE	15	70 min		200
Gamberale et al 1975	White spirit	14	4 x 30 min 1 x 60 min	4000 mg/m^3	2000 mg/m^3
Gamberale et al 1978	Xylene	23	70 min	300 ppm xylene and work on bicycle ergometer (100 W)	300
Savolainen et al 1979	Xylene	6	6 hours/day 2 weeks	100 ppm first day	200 ppm fifth day
Savolainen 1980	Xylene	10	4 hours		280
	Alcohol	10		1 ml/kg	0,5 ml/kg
	Combination	10	4 hours	1 ml/kg and 280 ppm	
Savolainen et al 1980	Xylene	8	6 hours	90	
Savolainen et al 1981	Xylene	9	4 hours		200
	Trichloroethane	9	4 hours		400
	Combination	9	4 hours		200 ppm Xylene and 400 ppm Trichloroethane (no interaction)

for the different solvents. As yet, no interactional effects of the non-additive type on behavioral performance have been demonstrated for combined exposure to two solvents or for the combination of exposure to one solvent and intake of alcohol or pharmaceuticals.

Many questions have been raised concerning the representativity and validity of the effects on behavioral performance observed in experimental laboratory investigations. Most of the criticism is directed toward aspects of experimental procedure and experimental control in general and will not be discussed here. Two issues do, however, seem to be more specifically relevant to the type of experimental inhalation studies referred to above. The first issue concerns the number and the choice of subjects participating in the experiments. For natural reasons only a very limited number of subjects can be studied in each experiment. Moreover, the subjects in most experiments have been chosen among young, healthy persons with high educational background rather than among solvent exposed workers. It is argued that this circumstance can have contributed to an increase in the number of experiments with negative results, thus leading to an overall underestimation of the potential capacity of solvent exposure to affect performance. The other issue concerns the ecological relevance of the experimental situation in general. Thus it is a basic limitation of the experimental approach that the exposure conditions which can be simulated in an exposure chamber are rarely if ever very representative of the conditions existing in the real work environment.

One way of studying the acute effects of solvent exposure and overcoming most of the shortcomings of the experimental approach is to conduct quasi-experimental studies directly in the field.

QUASI-EXPERIMENTAL FIELD STUDIES

Only a few attempts have been made to evaluate the acute effects of solvents using psychometric tests directly at the work site (Table 3). However, the results obtained in these investigations clearly demonstrate the narcotic effects of exposure to a number of organic solvents. Furthermore, these investigations give an interesting illustration of how the assessment of performance changes during the work shift may be of practical importance with regard

Table 3. Quasi-experimental field studies on solvent-induced behavioral performance changes.

AUTHOR	EXPOSURE	SUBJECTS	PERFORMANCE TESTS
Götell et al 1972	Styrene 150 ppm	17 exp. 17 contr. Mean age 28 yrs	Simple reaction time (SRT)
Gamberale et al 1976	Styrene 20 - 120 mm	106 exp. 36 contr. Mean age: 33 yrs	SRT
Binaschi et al 1976	Solvent mixtures 1.72 of TLV	8 exp.	Battery of 10 tests
Kjellberg et al 1979	Styrene 40 - 50 ppm	7 exp 7 contr. Mean age: 49 yrs	SRT, Mental arith, Bourdon
Cherry et al 1980 1981	Styrene 52 - 227 ppm	27 exp. 27 contr. Mean age: 23 - 26 yrs	SRT, Vigilance task, Digit symbol Digit span, Tapping
Anshelm Olson et al 1981	Solvent mixtures 1.0 - 1.5 of TLV	42 exp. Mean age: 31 yrs	SRT
Anshelm Olson 1982	Solvent mixtures 0.35 - 2.94 of TLV	47 exp. 47 contr. Mean age: 51 yrs	SRT, Bourdon, Benton Critical flicker fusion (CFF)

to the monitoring and prevention of hazards to health.

A common characteristic of the designs adopted in the field investigations reviewed below is that performance measures were collected both at the beginning and at the end of a work day. The rationale of this design is based on the underlying assumption that the effect of exposure during the day may manifest itself in an impairment of performance at the end of a work day, with a return to normal levels of performance by the beginning of the following work day, after approximately 16 hours free from exposure. If the action of exposure lasts long enough to bridge the exposure-free period between two consecutive work days, performance should also be negatively affected at the start of a work day.

Some features of these field studies are presented in Table 3.

In the study by Götell and coworkers (1972), the performance on a reaction time test of a group of workers exposed to an average of 150 ppm styrene was compared to that of a referent group of non-exposed workers of the same age. The mean reaction time of the exposed workers appeared to be affected both before and after a work day.

In the study by Gamberale and coworkers (1976), workers from four factories building fibre-glass boats, who were exposed to styrene in concentrations varying between 20 and 120 ppm, were compared with regard to reaction time with age-matched workers from two mechanical industries. The workers exposed to styrene had a longer reaction time, a greater deterioration of reaction time over time and a greater irregularity of performance on the test than the non-exposed workers both before and after a work day. However, these differences between the groups were more pronounced at the end of a work day.

The performance of a group of workers exposed to solvent mixtures at concentrations clearly exceeding the TLVs was studied by Binaschi and coworkers (1976). No control group was studied in this investigation. However, the effect of exposure manifested itself in a pronounced deterioration of performance after a work day. This deterioration of performance was observed in spite of the use of tests which in a control condition would have revealed a learning effect.

Kjellberg and coworkers (1979) were able to follow up a group of styrene exposed workers after the closing down of a fibre-glass boat factory. They found that the reaction time of the exposed was prolonged compared with that of a control group of non-exposed workers. This effect could still be observed four days after the cessation of exposure No effect was noticeable 35 days after the cessation of exposure.

In the investigations by Cherry and coworker (1980, 1981), reaction time was found to be prolonged among styrene exposed workers before the beginning of a work day. They also found that early morning urinary mandelic acid concentrations after two days without exposure correlated with reaction time measured on arrival at work.

In a longitudinal investigation by Anshelm Olson and

coworkers (1981) the reaction time of workers handling organic solvents and solvent based paints in the production of plastic-coated sheets was measured several times during a period of more than two years. The performance of the workers was found to improve over time as a result of changes in the ventilation system and in the working process, which had brought about a considerable reduction in solvent vapor concentration at the work place.

The effects of exposure to a mixture of organic solvents was also studied by Anshelm Olson (1982) among workers in the paint industry. In this study the exposed group performed less well than the control group on most of the behavioral tests used. The difference were evident for both morning and afternoon measurements. Particularly noticeable was a marked decrease in performance on a reaction time test over the course of the day for the workers exposed to the highest solvent levels.

Considering the results obtained in the investigations reviewed above, it seems reasonable to regard the assessment of behavioral performance as constituting a potential means for the monitoring and control of exposure to neurotoxic substances in a specific work environment. The detection of working conditions leading to CNS depression would clearly indicate the need for improvements in the hygienic quality of the environment under investigation and would, therefore, be a valuable complement to traditional environmental monitoring. The use of performance tests for this purpose does, however, make special demands on the procedure.

Since measurements of performance have to be carried out at different times during the day, circadian rhythms may constitute a confounding factor as important or even more important than learning or motivation. However, it does seem possible to minimize or control the effects of these confounding factors using operations of experimental control such as the use of an age-matched reference group and a balanced design. This was accomplished in some of the investigations reviewed above using a quasi-experimental study design of which a schematic representation is given in Table 4.

As shown in the table, half the workers in the exposed and control groups perform the tests before work the first

Table 4. Experimental design in field studies of acute effects of exposure to neurotoxic substances.

MEASUREMENT OCCASIONS		
BEFORE WORK	AFTER WORK	BEFORE WORK
Eg 1	Eg 1	-
	Eg 2	Eg 2
Cg 1	Cg 1	
	Cg 2	Cg 2

Eg 1, Eg 2 Exposed groups

Cg 1, Cg 2 Control groups of age-matched workers

time and after work the second time, whereas the other half perform the tests after work the first time and before work the second time. Balacing the study in this way, a single analysis of variance for each performance variable is sufficient to test the effects of exposure, of learning and of the circadian rhythm independently of each other and to test the possible effect of the interactions between these factors.

EPIDEMIOLOGICAL STUDIES

The evidence for the toxic effects of long term occupational exposure to low concentrations of solvent vapor is based largely on the results from epidemiological studies in which the behavioral performance of groups of exposed workers has been compared to that of reference groups. In order for such comparisons to be meaningful, the assumption of the comparability of the exposed groups to the reference groups as regards performance capacity prior to exposure must be valid. Unfortunately in retrospective studies of this type, there is no means of validating this assumption.

Therefore, no matter how rigorously the researchers perform the selection or matching of the groups, the comparability of the exposed to the reference group with regard to behavioral performance capacity will always be uncertain. A further contribution to this uncertainty is the fact that performance variables are usually highly correlated to background factors such as age, education, intelligence etc. These factors are known to have a selective impact on the individual's choice of job. Further difficulties in the interpretation of the outcome of this type of post factum studies are related to the nature of the independent variable. Thus in most cases there is no true knowledge of what the exposure has been, instead this is estimated from uncertain historical information.

Due to the above mentioned methodological limitation, the results of each specific study should be interpreted with caution. Possibly a better understanding of the state of affairs can be achieved by analysing and comparing the results of the different investigations.

Table 5 presents some features of a number of epidemiological studies conducted to assess the long term effects of solvent exposure. To facilitate the comparisons between the studies, an attempt has been made to classify the performance effects observed in each investigation on the basis of whether the tests used were supposed to assess primarily 1) memory functions, 2) other intellectual functions, 3) perceptual functions, 4) reaction time and 5) other psychomotor functions.

Two of the studies listed in Table 5 were conducted with groups of textile workers exposed to carbon disulphide (Hänninen, 1971; Putz-Anderson, 1983). The results of these two studies do not agree with regard to the effects on performance. In the study by Hänninen the intellectual and psychomotor functions of the exposed workers did appear to be affected, while no such effect could be observed in the investigation by Putz-Anderson and co-workers. One explanation for this difference in results may be that the exposure conditions seem to have been more severe in the study by Hänninen.

Groups of painters were investigated with extensive batteries of behavioral tests in two Finnish and two Swedish studies (Elofsson et al, 1980; Hane et al, 1977;

Table 5. Epidemiological studies on solvent-induced effects on behavioral performance.

AUTHOR	EXPOSURE	NO OF SUBJECTS EXPOSED	CONTROLS	MEASURED FUNCTIONS AFFECTED	NON AFFECTED
Hänninen 1971	Carbon disulphide	(50)+50	50	I, PM	M, P
Hänninen et al 1976	Solvent mixtures	100	101	M, I, P, PM	RT
Lindström et al 1976	Styrene	98	43	PM	M, I, P, RT
Hane et al 1977	Solvent mixtures	52	52	I, PM	M, P, RT
Knave et al 1978	Jet fuel	30	30	P, RT	M, PM
Elofsson et al 1980	Solvent mixtures	80	80	M, P, PM, RT	I
Cherry et al 1981	Methylene chloride	29	29		M, I, P
Anshem Olson 1982	Solvent mixtures	47	47	M, P, RT	(CFF)
Iregren 1982	Toluene	34	34	RT	M, I, P, PM
Lindström and Wickström 1983	Solvent mixtures	219	229	M, RT	I, PM
Putz-Anderson et al 1983	Carbon disulphide	131	167		M, I, RT, PM
Cherry et al 1984	Toluene	59	59	I (reading test)	M, P, RT, PM
	Solvent mixtures	42	42	RT	M, P, PM

M = memory functions; I = other intellectual functions; P = perceptual functions; RT = reaction time; PM = other psychomotor functions.

Hänninen et al, 1976; Lindström and Wickström, 1983). Statistically significant differences in the performance of various tests were obtained between exposed and non-exposed workers in all these studies. However, some differences between these studies should be especially noted. In the study by Hänninen et al (1976) the performance of the exposed group was clearly poorer than that of the reference group in all but one of the tests used. On the whole, the higher intellectual function seemed to be more affected by exposure than the psychomotor functions. In the study by Hane et al (1977) the differences in performance between the painters and the referent subjects were not so distinct. Only one test of intellectual ability and one of psychomotor functions revealed differences in performance. Moreover, in five out of 12 tests the painters actually did a little better than the referent workers. In the two more recent studies, the one by Elofsson et al (1980) and the one by Lindström and Wickström (1983),there was no manifest decrements in intellectual functions among the exposed workers. The differences observed concerned perceptual and psychomotor functions, particularly reaction time.

Recently, exposure to paint solvents was also investigated by Anshelm Olson (1982) among workers in the paint industry and by Cherry and coworkers (1984) among painters in a naval dockyard. Some features of the former study, which was designed for repeated measurements of performance, have already been presented. The only clearly consistent result of these two studies was that the exposed workers had longer reaction time than the non-exposed workers in the reference groups. These results cannot for certain be interpreted as due to long term exposure. Similar results were obtained by Iregren (1982) who examined a group of printers with a battery of tests of which only a reaction time task showed a decrement in performance.

In the remaining studies listed in table 5, either no effects of exposure were found or the effects were restricted to perceptual and psychomotor functions.

A few conclusions can be drawn from a comparative analysis of the results of the epidemiological investigations reviewed above. It is evident that at least some aspects of performance were found to be impaired in the great majority of the studies. This fact can hardly be attributed to chance. The argument that systematic errors in the choice of reference groups can have created spurious differences in performance in so many independently conducted investigations is not very convincing. It should be noted, however, as has been shown by Lindström and Wickström (1983) and by Cherry et al (1984) that differences in performance between exposed and non-exposed groups of workers do sometimes tend to disappear when the groups are matched with regard to the intellectual level of the workers before exposure. It is also interesting to observe that in both these studies the matching procedures did not affect the results of a test of simple reaction time showing a performance decrement among the exposed workers.

Clear evidence for the effects of long term exposure to solvents is also found in the results of two studies of styrene exposed workers (Gamberale et al, 1976; Lindström and Wickström, 1983). In these studies the duration of exposure was found to be negatively correlated to performance on certain tasks independent of the age of the exposed workers.

There seems to be no real evidence that the effects on behavioral performance due to long term solvent exposure are distinguishable from the effects which have been observed in laboratory experiments or in quasi-experimental field studies. In order to describe the existing differences between the acute and the long term manifestation of solvent exposure a more complex model is needed, which integrates the assessment of behavioral performance with that of other behavioral variables, i.e. neurobehavioral reactions, subjective symptoms, psychiatric observations etc.

CLINICAL STUDIES

Substantial evidence that long term exposure to organic solvents may produce neuropsychological impairments derives not only from the investigation of behavioral performance changes in exposed workers, but also from case reports, clinical studies and epidemiological studies of workers with disability pensions (for a review, see Hernberg, 1980). The solvent-induced clinical conditions have been given different names. In Finland and Sweden, for example, the condition is called "psychoorganic syndrome", while in Denmark and Norway the diagnos is labelled "presenile dementia". The non-specific manifestation of the toxicity of solvents, however, makes the individual clinical diagnosis very difficult.

In Sweden the diagnostic recognition of solvent induced disorders is mainly based on information concerning the exposure history of the worker/patient and on observations of his behavioral performance capacity, which are systematically carried out using a battery of psychometric tests (Hogstedt et al, 1980).

Essentially, the work of the clinical psychologist in this case is very similar to that performed in the assessment by psychometric methods of the cerebal functional status of a patient with suspected minor brain damage.

The psychometric methods used in clinical investigations in the different Scandinavian countries are very similar.They have been designed on the basis of the results obtained in the epidemiological investigations reviewed in the previous section. Further information useful in the development of the psychometric methods used in the clinical investigations derives from the analysis of results of groups of patients referred to occupational health clinics for suspected solvent-induced functional impairments (Ekberg and Hane, 1984).

CONCLUDING REMARKS

One of the conclusions which can be drawn from the present review is that many researchers have relied successfully on psychometric techniques to detect early signs of the toxicity of industrial solvents. Of course, other methods such as neurophysiological measurements and systematic observations of subjective symptoms have also been suitable for this purpose and have often been used together with the psychometric methods. However, it seems clear now that the measurement of behavioral performance has been demonstrated to possess more general applicability in human studies than other methods. Thus, the demonstration in experimental laboratory studies that behavioral performance changes do or do not occur at certain solvent concentrations is very important information for the establishment and evaluation of occupational exposure standards. The results of the application of psychometric techniques in quasi-experimental field studies have also been very promising. Probably the potentialities of these methods in the monitoring of hazards in the work environment have not yet been fully exploited. The use of the study design outlined in this paper, or the application of a similar strategy to collect measurements of performance directly at the working site appears to be practicable. Some behavioral tests suitable for this purpose are already available and others can be developed (Söderman et at, 1982; Iregren et al, 1985). To be suitable for use in the field a performance test must satisfy a number of requirements: the test must be sensitive, short, reliable, simple to administer, resistant to learning or training effects and, most important, the performance measures obtained must be related to early signs of functional disorders.

The evaluation of the psychometric methods with regard to the assessment of the long term effects of solvent on the central nervous system is more difficult. The difficulties, however, are not strictly related to specific characteristics of the psychometric methods. Instead, as discussed previously, the difficulties are mostly due to the limitations inherent in the epidemiological methods as such. With all likelihood a longitudinal approach with repeated measurements of performance over time would result in substantially more distinct results.

An important question, although difficult to answer, concerns the type of behavioral performance tests to be used in the study of the toxicity of solvents. Because of the complexity of the question no attempt will be made here to review this argument in detail. Only a few comments on this topic will be made in these final remarks.

To begin with, it is not possible to give this quesion a fully valid answer because this presupposes an understanding of the neurophysiological and neuropsychological mechanisms (affected by the solvents),knowledge which we do not yet possess. Moreover, it is important to keep in mind that, independent of their use, performance tests are supposed to provide measurements of the activity of underlying functional process of which in fact very little is known. For this reason it is often very difficult to define what is actually measured by different performance tests. Even the relatively simple classification of performance tests carried out in this paper (table 5) must be considered arbitrary. In fact, although a distinction is commonly drawn between different types of mental functions which can be measured by behavioral performance tests, this is very difficult to maintain completely. Thus, perception, decision, knowledge, judgment and coordinated overt activity by the hands, organs of speech or other effectors are involved in all performance tests. Usually the tests are defined and classified according to which mental function is supposed to play the most essential part in performance. However, it should be observed that an impairment of performance need not necessarily be related to the mental function the test is supposed to assess.

Until the mechanisms of solvent toxicity are better understood, it is advisable to base the choice of behavioral tests on strictly empirical considerations. There is no doubt that the likelihood of detecting adverse effects of solvent exposure will be enhanced if the performance tests used involve sensory-motor, vigilance and perceptual functions rather than functions related to intelligence. Performance tests of intellectual functions could possibly be used for matching procedures. There is no doubt that psychometric methods can and need to be improved. A promising development consists in the computerization of test procedures. Computerized batteries of performance tests have already been used in experimental laboratory studies (Anshelm Olson et al, 1985; Iregren et al, 1985) and found to have considerable advantages as compared to paper and pencil tests. In Sweden a collaboration is in progress between the Research Department of the NBOSH and some of the occupational health clinics with the task of evaluating the application of a computerized battery of psychometric tests for use in clinical examinations (Iregren et al,1985; Almkvist et al, 1985).

REFERENCES

Almkvist O, Iregren A, Wallén M, Åslund U (1985). Evaluation of some computer-administrated behavioral tests for diagnosing CNS dysfunction after long-term solvent exposure: a study in progress. In: Neurobehavioral methods in occupational and environmental health. Second international symposium,Copenhagen 6-9 August 1985. WHO Regional office for Europe, Copenhagen.

Andersen I, Lundqvist GR, Mölhave L, Find Pedersen O, Proctor DF, Vaeth M, Wyon DP (1983). Human response to controlled levels of toluene in six-hour exposures. Scand J Work Environ Health 9:404-418.

Anshelm Olson B, Gamberale F, Grönqvist B (1981). Reaction time changes among steel workers. A longitudinal study. Int Arch Occup Environ Health 48:211-218.

Anshelm Olson B (1982). Effects of organic solvents on behavioral performance of workers in the paint industry. Neurobehav Toxicol Teratol 4:703-708.

Anshelm Olson B, Gamberale F, Iregren A (1985). Coexposure to toluene and p-xylene in man: central nervous functions. Br J Ind Med 42:17-22.

Binaschi S, Gazzaniga G, Crovato E (1976). Behavioural toxicology in the evaluation of the effects of solvent mixtures. In: M Horvath ed. Adverse effects of environmental chemicals and psychotropic drugs, vol. 2. Elsevier Scientific Publishing Co. Amsterdam 1976 pp 91-98.

Cherry N, Waldron HA, Wells GG, Wilkinson RT, Wilson HK, Jones S (1980). An investigation of acute behavioural effects of styrene on factory workers. Br J Ind Med 37:234-240.

Cherry N, Rodgers B, Venables H, Waldron HA, Wells GG (1981). Acute behaviour effects of styrene exposure: a further analysis. Br J Ind Med 38:346-350.

Cherry N, Venables H, Waldron HA, Wells GG (1981). Some observation on workers exposed to methylene chloride. Br J Ind Med 38:351-355.

Cherry N, Johnston JD, Venables H, Waldron HA (1983). The effects of toluene and alcohol on psychomotor performance. Ergonomics 26:1081-1087.

Cherry N, Venables H, Waldron H (1984). British studies on the neuropsychological effects of solvent exposure. Scand J Work Environ Health 10, Suppl.1:10-12.

Dick RB, Setzer JV, Wait R, Hayden MB, Taylor BJ, Tolos B, Putz-Andersen V (1984). Effects of acute exposure of toluene and methyl ethyl ketone on psychomotor performance. Int Arch Occup Environ Health 54:91-109.

Di Vincenzo GD, Yanno FJ, Astill BD (1972). Human and canine exposures to methylene chloride vapor. Am Ind Hyg Assoc J 33:125.

Ekberg K, Hane M (1984). Test battery for investigating functional disorders - The TUFF battery. Scand J Work Environ Health, 10 Suppl.1:14-17.

Elofsson S-A, Gamberale F, Hindmarsh T, Iregren A, Isaksson A, Johnsson I, Knave B, Lydahl E, Mindus P, Persson HE, Philipson B, Steby M, Struwe G, Söderman E, Wennberg A, Widen L (1980). Exposure to organic solvents. A cross--sectional epidemiological investigation on occupationally exposed car and industrial spray painters with special reference to the nervous system. Scand J Work Environ Health 6:239-273.

Ettema JH, Kleerekoper L, Duba WC (1975). Study of mental stress during short-term inhalation of trichloroethylene. Staub-Reinh Luft 35:409-410.

Ferguson RK, Vernon RJ (1970). Trichloroethylene in combination with CNS drugs (effects on visual motor tests). Arch Environ Health 20:462-467.

Gamberale F, Hultengren M (1972). Toluene exposure: II. Psychophysiological functions. Work Environ Health 9:131-139.

Gamberale F, Hultengren M (1973). Methylchloroform exposure II. Psychophysiological functions. Work Environ Health 10:82-92.

Gamberale F, Hultengren M (1974). Exposure to styrene: II. Psychological functions. Work Environ Health 11:86-93.

Gamberale F, Annwall G, Hultengren M (1975). Exposure to white spirit: II. Psychological functions. Scand J Work Environ Health 1:31-39.

Gamberale F, Annwall G, Hultengren M (1975). Exposure to methylene chloride: II. Psychological functions. Scand J Work Environ Health 2:95-103.

Gamberale F, Annwall G, Anshelm Olson B (1976). Exposure to trichloroethylene: III. Psychological functions. Scand J Work Environ Health 4:220-224.

Gamberale F, Lisper H O, Anshelm Olson B (1976).The effect of styrene vapour on the reaction time of workers in the plastic boat industries. In: M Horvath ed. Adverse effects of environmental chemical and psychotropic drugs, vol. 2. Elsevier Scientific Publishing Co. Amsterdam pp 135-148.

Gamberale F, Annwall G, Hultengren M (1978). Exposure to xylene and ethylbenzene: III. Effects on central nervous functions. Scand J Work Environ Health 4:204-211.

Gamberale F, (1985). Use of behavioral performance tests in the assessment of solvent toxicity. Scand J Work Environ Health 11 Suppl 1(in press).

Götell P, Axelson O, Lindelöf B (1972). Field studies on human styrene exposure. Work Environ Health 9:76-83.

Hane M, Axelson O, Blume J, Hogstedt C, Sundell L, Ydreborg B (1977). Psychological function changes among house painters. Scand J Work Environ Health 3:91-99.

Hernberg S (1980). Neurotoxic effects of long-term exposure to organic hydrocarbon solvents. Epidemiologic aspects. In: Holmstedt B, Lauwerys R, Mercier M, Roberfroid M eds. Mechanisms of toxicity and hazard evaluation. Elsevier

Amsterdam. pp 307-317.

Hogstedt C, Hane M, Axelson O (1980).Diagnostic and health care aspects of workers exposed to solvents. In: Zenz, C ed. Developments in occupational medicine. Year boook medical publishers, Inc. Chicago, London. pp 249-258.

Hänninen H (1971). Psychological picture of manifest and latent carbon disulphide poisoning. Br J Ind Med 28:374-381.

Hänninen H, Eskelinen L, Husman K, Nurminen M (1976). Behavioral effects of long-term exposure to a mixture of organic solvents. Scand J Work Environ Health 4:240-255.

Iregren A (1982). Effects on psychological test performance of workers exposed to a single solvent (toluene) - A comparison with effects of exposure to a mixture of organic solvents. Neurobehav Toxicol Teratol 4:695-701.

Iregren A, Åkerstedt T, Anshelm Olson B, Gamberale F, Wallén M (1985). Experimental exposure to toluene in combination with ethanol intake: psychophysiological and behavioral effects (to be published).

Iregren A, Gamberale F, Kjellberg A (1985). A microcomputer based behavioral testing system. In: Neurobehavioural methods in occupational and environmental health. Second international symposium, Copenhagen 6-9 August 1985. Who Regional office for Europe, Copenhagen.

Kjellberg A, Wigaeus E, Engström J, Åstrand I, Ljungquist E (1979). Long-term effects of styrene exposure in a polyester plant (In Swedish). Arbete och Hälsa 18:1-25.

Knave B, Anshelm Olson B, Elofsson S, Gamberale F, Isaksson A, Mindus P, Persson HE, Struwe G, Wennberg A, Westerholm P (1978). Long-term exposure to jet fuel. II. A cross--sectional epidemiologic investigation on occupationally exposed industrial workers with special reference to the nervous system. Scand J Work Environ Health 4:19-45.

Konietzko H, Elster I, Sayer M, Weichardt M (1975). Zentralnervose storungen durch trichlorethylen. Staub-Reinhalt Luft 35:240-241.

Lindström K, Härkönen H, Hernberg S (1976). Disturbances in psychological functions of workers occupationally exposed to styrene. Scand J Work Environ Health 2:129-139.

Lindström K, Wickström G (1983). Psychological function changes among maintenance house painters exposed to low levels of organic solvent mixtures. Acta Psychiat Scand 67, Suppl. 303:81-91.

Nakaaki K, Onishi N, Iida H, Kimotsuki K, Fikabori S, Morikiyo Y (1973). An experimental study on the effect of exposure to trichloroethylene vapor in man. Japanese

Science Labour 44:449-463.

Putz VR, Johnson B, Setzer J (1979). A comparative study of the effects of carbon monoxide and methylene chloride on human performance. J Environ Pathol Toxicol 2:97-112.

Putz-Anderson V, Setzer JV, Croxton JS, Phipps FC (1981). Methyl chloride and diazepam effects on performance. Scand J Work Environ Health 7:8-13.

Putz-Anderson V, Albright B, Lee S, Johnson B, Chrislip D, Taylor B, Brightwell S, Dickerson N, Culver M, Zentmeyer D, Smith P (1983). A behavioral examination of workers exposed to carbon disulphide. NeuroToxicology 4:67-78.

Salvini M, Binaschi S, Riva M (1971). Evaluation of the psychophysiological functions in human exposed to the threshold limit value of 1,1,1-trichloroethane. Br J Ind Med 28:286-292.

Salvini M, Binaschi S, Riva M (1971). Evaluation of the psychophysiological functions in humans exposed to trichloroethylene. Br J Ind Med 28:293-295.

Savolainen K, Riihimäki V, Linnoila M (1979). Effects of short-term xylene exposure on psychophysiological functions in man. Int Arch Occup Environ Health 44:201-211.

Savolainen K, Riihimäki V, Seppäläinen AM, Linnoila M (1980). Effects of short-term m-xylene exposure and physical exercise on the central nervous system. Int Arch Occup Environ Health 45:05-121.

Savolainen K (1980). Combined effects of xylene and alcohol on the central nervous system. Acta Pharmacol Toxicol 46: 366-372.

Savolainen K, Riihimäki V, Laine A, Kekoni J (1981). Short-term exposure of human subjects to m-xylene and 1,1,1-trichloroethane. Int Arch Occup Environ Health 49:89-98.

Stewart RD, Dodd HC, Baretta ED, Schaffer AW (1968). Human exposure to styrene vapor. Arch Environ Heal 16:656-662.

Stewart RD, Gay HH, Schaffer AW, Early DS, Rowe VK (1969). Experimental human exposure to methyl chloroform vapor. Arch Environ Health 19:467-472.

Stewart RD, Baretta ED, Dodd HC, Torkelson TR (1970). Experimental human exposure to tetrachloroethylene. Arch Environ Health 20:225-229.

Stewart RD, Hake CL, Le Brun AI, Kalbfleisch JH, Newton PE, Peterson JE, Cohen H, Strube R, Busch KA (1974). Effects of trichloroethylene on behavioral performance capabilities. In: V Xintaras, BL Johnson, I Groot eds. D H E W Publication no (NIOSH) 74-126. U S Department of Health, Education and Welfare, Washington D C, pp 96-126.

Stewart RD (1975). Toluene: Development of a biological standard for the industrial worker by breath analysis. National Institute for Occupational Safety and Health. Cincinnati, (NiOSH Contract Report No 99-72-84).

Stewart RD, Hake CL, Wu A (1977). Effects of perchlorethylene drug interaction on behavior and neurological function. USDHEW (NIOSH) Publication No. 77-191. Washington.

Söderman E, Kjellberg A, Anshelm Olson B, Iregren A (1982). Standardisation of a simple reaction time test for use in behavioral toxicology research (In Swedish with English summary). National Board of Occupational Safety and Health, Sweden. Undersökningsrapport no 27.

Vernon RJ, Ferguson RK (1969). Effects of trichloroethylene on visual-motor performance. Arch Environ Health 18:894.

Windemüller FBJ, Ettema JH (1978). Effects of combined exposure to trichloroethylene and alcohol on mental capacity. Int Arch Occup Environ Health 41:77-85.

Winneke G (1974). Behavioral effects of methylene chloride and carbon monoxide as assessed by sensory and psychomotor performance. In: C. Xintaras, BL Johnson, I de Groot eds. Behavioral Toxicology, U S Government Printing Office, Washington, pp 130-144.

Winneke G (1982). Acute behavioral effects of exposure to some organic solvents - psychophysiological aspects. Acta Neurolog Scand Suppl 92, 66:117-129.

Safety and Health Aspects of Organic Solvents, pages 225–236

NEUROBEHAVIORAL ASSESSMENT OF LONG-TERM SOLVENT EFFECTS ON MAN

Helena Hänninen

Department of Psychology, Institute of Occupational Health, Haartmaninkatu 1, SF-00290 Helsinki, Finland

INTRODUCTION

The neurotoxic effects of organic solvents often manifest themselves as behavioral dysfunctions. In the case of long-term exposures, the incipient neurotoxic effect produces quite subtle and often transitory deficits in behaviour. In the earliest stage, performance may be defective only in situations where the person's capacity to compensate for minor dysfunctions is lowered by some internal or external factor(s), such as temporary fatigue or nervousness, or a distracting environment. With a more advanced toxic effect, the behavioral dysfunction is more permanent, involves a larger variety of behavioral functions, and leads to more severe performance decrements. Psychological test methods are used to study the early neurotoxic effects in exposed worker groups, and also to detect toxic effects in individuals with suspected organic solvent intoxication.

In the following discussion I shall first describe the test methods that are used for this purpose, then present several study approaches and give some examples of the empirical studies. Lastly, I shall speak about the clinical neuropsychological evaluation of solvent intoxications.

TEST METHODS

Because the effects of long-term toxic exposure often involve several psychological functions, we need a versatile set of tests to obtain a valid picture of the neurobehavioral dysfunctions. (Hänninen and Lindström, 1979)

A comprehensive test battery includes tests for:

- motor speed,
- hand steadiness,
- perceptual speed,
- coordination of the perceptual and motor functions (reaction speed, eye-hand coordination, manual dexterity),
- verbal intelligence (vocabulary, verbal concept formation, verbal fluency),
- nonverbal intelligence (visual organization, visuoconstructive ability),
- auditive (verbal) memory and learning,
- visual memory,
- emotional reactions.

Such a comprehensive test battery is necessary when the test battery is used for an individual diagnosis of neurotoxic damage. When used for that purpose the test battery must also be standardized, which means that it is administered in a consistent manner from one occasion to another and the test results are converted into standard scores. The standard scores are needed for both inter-individual and intra-individual comparisons of the test performances.

In epidemiological studies we must often apply a more limited set of tests. Tests for motor functions, eye-hand coordination, visuoconstructive ability, and auditive and visual memory have proven to be the most useful ones. The use of the verbal intelligence tests for detecting neurobehavioral dysfunctions is more limited. On the other hand, vocabulary tests, which are resistant to recent brain damage, are often applied to get an estimation of the premorbid intellectual level.

The toxic induced emotional changes that occur may appear in the forms of decreased emotional reactivity, or emotional lability. The affected person may suffer from recurrent spells of dysphoric moods, or from more sustained periods of tension, anxiety, irritability, etc. The rather severe depressions seen among those exposed to CS_2 are seldom seen in those exposed to other solvents. In clinical examinations these changes are assessed by a clinical interview and by various personality tests. In epidemiological studies personality and mood inventories can be applied for this purpose.

STUDY APPROACHES FOR INVESTIGATING THE EFFECTS OF LONG-TERM EXPOSURES

Early effects in exposed worker groups can be studied (a) by cross-sectional comparisons of the behavioral functions of exposed workers with those of matched referents, (b) by analysis of the relationship between exposure level and behavioral dysfunction within an exposed worker group, or (c) by prospective follow-up of the behavioral functions of exposed workers.

Approaches used for studying neurobehavioral dysfunctions among patients with solvent intoxication have included comparison with normal worker groups and comparison with other neurological patient groups.

Neurotoxic effects of solvents have mostly been investigated by cross-sectional studies. However, in addition to the cross-sectional data, several studies have also included retrospective data concerning the previous exposure levels, and some studies have also used retrospective data concerning previous (pre-employment) performance levels of the subjects. (E.g. Hänninen et al., 1976; Lindström and Wickström, 1983)

Comparison of the performances of exposed subjects and their referents has been an usual approach; it is the only possible method when no sufficient and valid information about individual exposure levels is available. In several studies, however, both approaches a and b were applied: the data analyses included both comparisons of the mean performances in the exposed group with those of a reference group, and a correlational analysis of the relationship between the exposure level and the behavioral dysfunction within the exposed group. Sometimes only group differences were found, but the performance decrements did not show a statistically significant correlation with the level or duration of the exposure. At other times the result revealed an exposure-effect relationship within the exposed group though the group comparison failed to show any difference. Crucial factors in these kinds of studies are the matching of the reference subject group with the exposed group in regard to possible positive and negative confounders, and the reliability of the present and past exposure data. Difficulties in matching and the nonavailability of sufficient exposure data are especially problematic in studies concerning neurobehavioral dysfunctions in patient groups.

Most of the behavioral studies have dealt with the exposure-effect and not with the exposure-response relationship. This means that the results were analyzed by comparing the mean performances in exposed and nonexposed worker groups, or by calculating correlation coefficients between test scores and the exposure indices, instead of comparing the prevalence of abnormal test results (response) at various levels of exposure, or among the exposed and nonexposed workers. The exposure-effect approach is often considered more adequate in respect to behavioral dysfunctions because of the large inter-individual variation of the psychological test scores. Besides, in the case of the very subtle early effects, we may find a difference at the group level, even when all or almost all of the performances are still within the range of normality (depending, of course, on how the range of normality was defined which is always a problematic issue in neurobehavioral studies).

But there are also some studies that dealt with the exposure-response relationship, and could demonstrate an increase of abnormal scores by the increasing level or duration of exposure. This kind of information is indeed very helpful, for example in evaluating health based standards for occupational exposures.

REVIEW OF SOME EMPIRICAL STUDIES

Behavioral effects of solvent mixtures used in painting have been documented by four Scandinavian research groups, each of which used a versatile standardized test battery (unfortunately, not quite the same).

In the car painter study conducted at our Institute (Hänninen et al., 1976) no measures or estimations of the individual exposure levels were available; the average level of exposure corresponded to one third of the hygienic standard for mixed exposures used in Finland at that time. The conclusions relied on group comparisons which were made both for the entire groups of exposed subjects and the referents, and for smaller subgroups which we were able to match in regard to their initial performance levels. For the matching we used test results that were obtained during the subjects' previous military service.

Figure 1 shows the mean performance of the painters in 14 test variables in standardized scores for the whole sample and for the matched group.

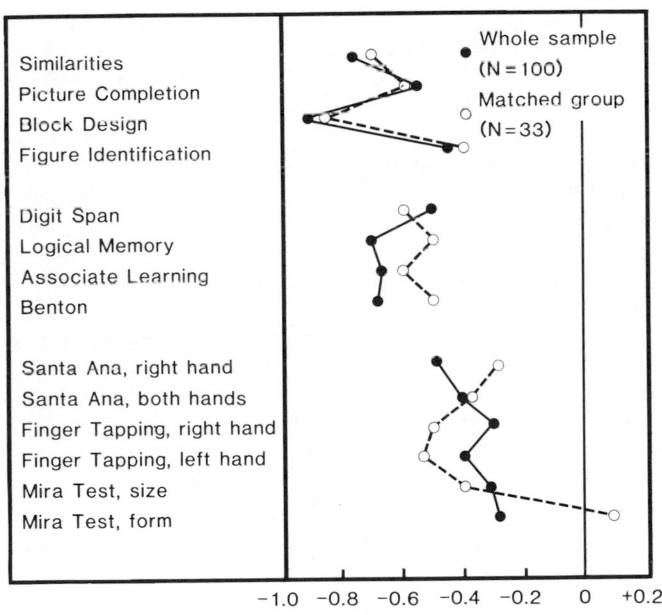

Figure 1. Mean performances of the exposed subjects (N=100 and N=33) in standardized scores. (The corresponding reference groups are used as standards.)

The painters' performance was inferior to that of their referents in several tests. Group differences were greatest in tests of verbal and visual intelligence, and in memory tests. Less difference was seen in tests that measure perceptual speed and motor abilities. The pattern of group differences was rather similar though not identical for total groups and for the matched subgroups. Our findings can be compared with those of a Swedish research group which conducted a similar type of study investigating car and industrial spray painters exposed to solvent mixtures (Elofsson et al, 1980). The level of exposure was approximately the same as in our study. The authors found disturbances in short term memory, perceptual speed and motor performances, including reaction speed. Contrary to the Finnish study the effects were more apparent in the perceptual and motor tasks than in the domain of intellectual performance.

The effects found among house maintenance painters have been more subtle. Hane and co-workers found slight group differences in the tests for visuoconstructive ability and motor coordination (Hane et al, 1977). A Finnish study (Lindström & Wickström, 1983) examined 219 house painters exposed on the average to 40 ppm white spirit for an average of 22 years. A subgroup of painters was again matched with nonexposed referents on age and initial intelligence level, measured during their military service. Comparison of the two groups showed differences in visual memory and reaction speed. In this study the total and average exposures to solvents during each subject's lifetime as well as during the last five years and 12 months were calculated or estimated. Using these estimates, a statistically significant dose-response relationship was found for attention, perceptual-motor speed, and verbal concept formation.

Of the few solvents that appear as the only exposure agent in certain occupations, styrene has been studied the most intensively.

In a Finnish study on styrene (Lindström et al, 1976; Härkönen et al, 1978) the urinary mandelic acid concentrations, measured during a five-week period before the examination, were used as a measure for the individual exposure levels. Comparison between the exposed and referent subjects revealed only mariginal differences between the groups, but the intragroup analyses showed statistically significant relationships between the mandelic acid concentrations and tests requiring visuomotor accuracy and control of hand movements. An increase of abnormal scores in these tests was found at the mandelic acid concentration level which corresponds to exposure levels of around 50 ppm as a mean daily exposure.

In a similar study, Mutti and co-workers found, at about the same exposure level, an increase not only of visuomotor and psychomotor impairments but also of impairments in visuoconstructive tests and in tests for verbal memory and learning. A statistically significant increase of verbal learning impairments was, in fact, found even at a considerably lower level of exposure (Mutti et al, 1984).

The authors also analyzed the predictive validity of the test battery in its entirety by comparing the prevalences of at least one, two and three abnormal scores in subgroups representing different intensities of exposure. The obtained dose-response curves can be seen in figure 2.

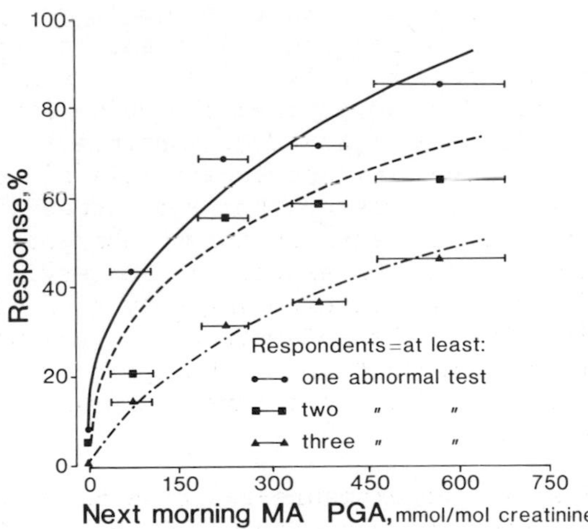

Figure 2. Dose-response relationships between intensity of styrene exposure, as measured by means of the urinary excretion of styrene metabolites and neuropsychological performance. (Abnormal tests were those poorer than the mean $\pm$ 2 SD of reference values.)

When the least restrictive criteria for a "response" was adopted, i.e. the presence of at least one abnormal score, 85 % of the most heavily exposed workers showed a response, as compared with 8 % of the control subjects and 43 % of the lowest exposure group. When the response was defined as "at least three abnormal scores" none of the controls showed a response. At the lowest exposure level the response prevalence was 14 %, in the most heavily exposed group, 46 %.

We have made similar kinds of analyses in a current study (Hänninen et al, in preparation), where the subject group comprised 205 workers, 178 of whom were exposed to organic solvents in different branches of industry. Most of them were exposed to solvent mixtures, but some also to single solvents, such as toluene and 1,1,1-trichlorethane.

The level of exposure was measured for each individual, and converted to a percentage of the hygienic limit value.

The aim of this still ongoing study is to evaluate the use and validity of a short test battery (seven tests yielding in all ten test scores) as a screening method in the health surveillance of workers exposed to organic solvents. To avoid too many false alarms, we wanted to have a method that screens only those with rather excessive impairments. Therefore we applied the quite strict screening criteria of at least four abnormal test scores. Those with two to three low scores were classified as borderline cases. We compared the prevalences of the abnormal and borderline cases in three subgroups representing different levels of exposure: the very low one (the mean daily exposure level being only 10 % of the hygienic limit value), the moderate one (where the mean daily exposure was around 40 % of the hygienic limit), and a high one (where the daily exposures were around the hygienic standard).

Figure 3 shows two exposure-response curves: the one obtained by defining the response as at least two abnormal scores, and the other obtained by defining it as at least four abnormal values. In this study our nonexposed workers provided us with more false positive cases than the group with a very low-level solvent exposure. Within the exposed subgroups a clear exposure-response relationship was seen.

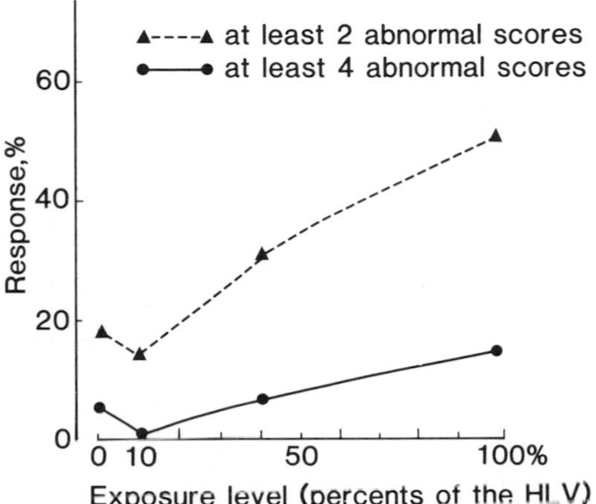

Figure 3. Dose-responce relationships between intensity of solvent exposure and neuropsychological performance.

NEUROPSYCHOLOGICAL EVALUATION OF SOLVENT INTOXICATIONS

As to the patients with organic solvent intoxication, the main impairments are generally seen in motor functions: in the speed as well as in the control of hand movements, and particularly in the manual dexterity tests. Very often the memory tests also show moderate to severe impairments, though it is possible to see quite intact memory test performances in spite of marked subjective memory difficulties. The occurrence of intellectual deficits is very variable across individual cases; most often they are seen in visuoconstructive tasks.

A clinical evaluation of the individual test results always includes an estimation of the premorbid level. Sources for this information are:

1. Interview of the patient, concerning school progress (particularly language skills and mathematics), progress and success at work, type of previous and present free-time activities, and perceived changes in mastering cognitive and motor skills.

2. Psychological test results, considering especially the actual performance in verbal tests that measure well learned abilities, and the level of best actual test performances.

The estimate must be based on multiple criteria because both the retrospective data based on the interview and the actual measurements of the cognitive abilities as such may be unreliable indicators of the real premorbid performance level.

In a clinical evaluation conclusions are drawn about the abnormality of the test results, and about the probability of different etiologies for the abnormal results. When conclusions about the abnormality of the results are drawn, the following factors are taken into account:

- estimation of the initial level of performances,
- age of the subject,
- educational and occupational background.

Situational factors that may account for a lowered performance level are also considered.

Evaluation of the etiological possibilities includes an evaluation of how the abnormal findings fit with our knowledge about the neurotoxic effects of solvents, and an evaluation of the probability of other etiological factors, either as alternates or as additional factors in a multifactorial etiology.

USE OF NEUROBEHAVIORAL METHODS IN THE PREVENTION OF NEUROTOXIC DISEASES CAUSED BY ORGANIC SOLVENTS

To sum up, neurobehavioral test methods have proven to be useful in epidemiological research for evaluating neurotoxic hazards in exposed worker groups, and for defining exposure-effect or exposure-response relationships as a basis of standard setting. These methods have also proven useful in clinical evaluation of solvent intoxication, whereas there is less documentation concerning their utilization in health surveillance of workers exposed to organic solvents.

The latter kind of practical application would, indeed, be helpful to find the sensitive individuals who show adverse effects even at those low levels of exposure that are usually considered as relatively safe. The use of neurobehavioral tests in health surveillance might be especially desirable in problematic exposure conditions as, for instance, in multicomponent exposures with an unknown synergism of toxicity, in highly varying solvent exposure, or in other situations where the health hazard due to organic solvents or solvent mixtures is difficult to estimate.

The methods used in epidemiological studies and in clinical practice need further development and modification to suit this kind of practical application. Baseline measurements made at the beginning of exposure would increase the validity and sensitivity of neurobehavioral methods when applied for identifying individuals with early effects.

Of course, behavioral functions are sensitive to other factors besides toxic exposure, and therefore they are apt to produce positive findings even when the baseline performance levels are known. These other factors must be ruled out by a more thorough examination in the case of positive findings. However, the unspecificity of the behavioral impairments caused by organic solvents is relative. With adequate neurobehavioral methods and adequate psychodiagnostic knowledge it is possible to differentiate toxic impairments from, say, disturbances caused by acute anxiety or by chronic neuroticism, though it is more difficult to differentiate them from certain types of diffuse brain damage.

REFERENCES

Elofsson SA, Gamberale F, Hindmarsch T, Iregren A, Isaksson A, Johnsson I, Knave B, Lydahl E, Mindus P, Persson HE, Philipson B, Steby M, Struwe G, Söderman E, Wennberg A, Widen L (1980). Exposure to organic solvents. A cross-sectional epidemiologic investigation on occupationally exposed car and industrial spray painters with special reference to the nervous system. Scand. J. Work Environ. & Health 6:239-273.

Hane M, Axelson O, Blume J, Hogstedt C, Sundell L, Ydreborg B (1977). Psychological function changes among house painters. Scand. J. Work Environ. & Health 3:91-99.

Hänninen H, Eskelinen L, Husman K, Nurminen M (1976). Behavioral effects of long-term exposure to a mixture of organic solvents. Scand. J. Work Environ. & Health 2:240-255.

Hänninen H, Lindström K (1979). Behavioral test battery for toxicopsychological studies. Used at the Institute of Occupational Health in Helsinki. Second revised edition. Reviews 1. Institute of Occupational Health, Helsinki.

Hänninen H, Rantala K, Tuominen E. Psychological methods in the health surveillance of workers exposed to solvents. In preparation.

Härkönen H, Lindström K, Seppäläinen AM, Asp S, Hernberg S (1978). Exposure-response relationship between styrene exposure and central nervous function. Scand. J. Work Environ. & Health 4:53-59.

Lindström K, Härkönen H, Hernberg S (1976). Disturbances in psychological functions of workers occupationally exposed to styrene. Scand. J. Work Environ. & Health 3:129-139.

Lindström K, Wickström G (1983). Psychological function changes among maintenance house painters exposed to low levels of organic solvent mixtures. Acta psychiat. Scand. 67:Suppl. 303, 81-91.

Mutti A, Mazzucchi A, Rustichelli P, Graziano F, Giuseppe A, Franchini I (1984). Exposure-effect and exposure-response relationships between occupational exposure to styrene and neuropsychological functions. Am. J. of Ind. Med. 5:275-286.

Safety and Health Aspects of Organic Solvents, pages 237–246

NEUROTOXICOLOGICAL AND NEUROPHYSIOLOGICAL EFFECTS OF SOLVENTS WITH MAINLY CENTRAL ACTION

Arne Wennberg

Department of Occupational Neuromedicine,
Research Division,
National Board of Occupational Safety and Health,
171 84 SOLNA, Sweden

Introduction

In high doses, probably all organic solvents have a toxic effect upon the central nervous system. The acute, narcotic effects have been known for more than 100 years and this qualification of the solvents has been used for surgical anaesthesia since the end of the last century. The reason for this general effect of the organic solvents is their property of solving fat or fat-similar tissues such as nervous tissue.

The nervous system is built up by highly specialized cells, the nerve cells or the neurons, and supporting cells of different kinds, the glial cells. The number of neurons in the human brain has been estimated to around 20 billions and they are all present at birth. There in no renewal of the neurons during the life, and in this respect the nervous system differs from other organs in the body, such as skin, liver, lung etc., where new cells are produced during the entire life. We must thus be very careful with our brain cells, because if some of them are destroyed early in the life, we have to manage the rest of the life with a reduced number of nerve cells. We normally loose a certain number of nerve cells during the life, and this procedure starts quite early, around the age of twenty years. However, if this reduction for some reason is increased, the situation in the later part of the life might be critical. One way of increasing the reduction of nerve cells might be such as heavy or long-term exposure to neurotoxic compounds as organic solvents.

The function of a neuron is very complex. It is schematically presented in Figure 1.

FIGURE 1. THE NEURONE

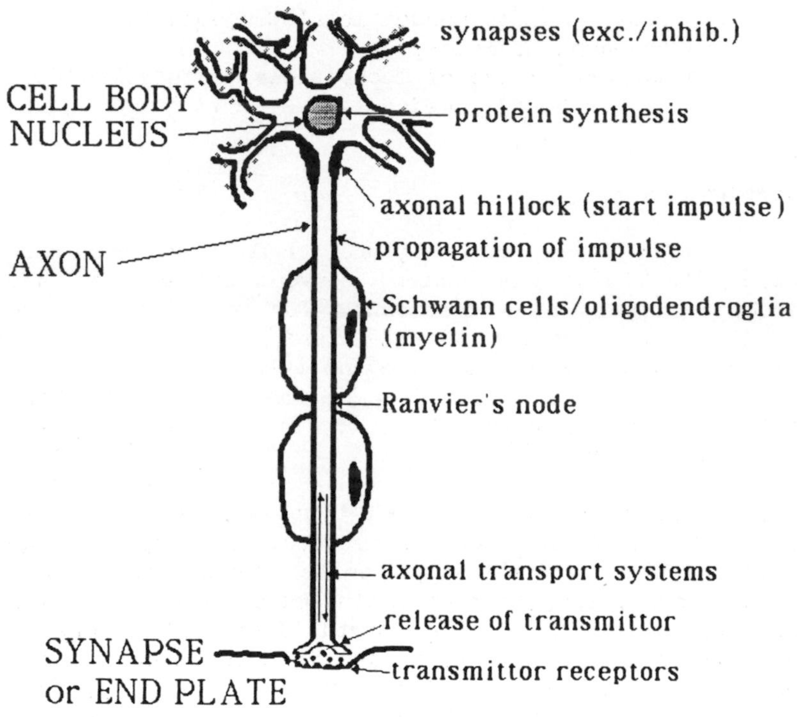

All the different functions given in the figure must be intact in order to maintain a normal function of the neuron. This also includes the myelin, that encircles the axons. The myelin is produced by Schwann cells in the peripheral nervous system and by the oligodendroglial cells. It should also be noted that the synaptic transmission of signals from one neuron to another involves a chemical compound, a neurotransmitter. There are different transmitters in different neurons. Thus a toxic compound that interferes with one particular transmittor has only effect upon those neurons using that transmittor. As examples of neurotransmittors, the following could be mentioned: cathecolamines, serotonine, acetylcholine, gamma-amino-butyric acid (GABA). The total number of transmittors in the nervous system is not definitively established yet. There are probably still a number of them to be discovered.

The organic solvents have toxic influence upon the nervous system, but it is not definitively established which one is disturbed of all different neuronic functions (cf Fig 1). It might even be several of them. The myelin is a fatty compound and the neuron membranes contains fat-similar molecules. It may therefore be reasonable to assume that these two structures are target-points for the solvents. It may also be reasonable to assume that solvents can interfere chemically with metabolic processes for producing or breaking down some of the neurotransmittors.

During the last decade, neurotoxic effects of organic solvents has been extensively studied, in particular in the Nordic countries. Interest has been focussed both upon acute and cronic effects and the studies has been experimental as well as epidemiological, both on animals and man. An important question has of course been whether long-term exposure for low concentrations (as often exists in current occupational environments) could cause damage to the nervous system. It was shown to be the case in several studies from our countries but there was initially great scepticism to those results. However, the situation to-day is different and I believe that a majority now accepts toxic effects upon the central nervous system caused by long-term occupational exposure to organic solvents, also when the exposure level is around present TLV:s.

Organic solvents with toxic effects upon central nervous system

Spencer and Schaumburg (1985) have raised criteria for neurotoxicity. They say that the following three questions must be answered:

(1) Does the substance or mixture produce a consistent pattern of neurological dysfunction in humans?

(2) Can this entity be induced in animals under comparable exposure conditions?

(3) Are there reproducible lesions in the nervous system or special sense organs of exposed humans and/or animals, and do these abnormalities satisfactorily account for the neurobehavioral dysfunction?

Failure to satisfy any of these criteria leaves, according to Spencer and Schaumburg, room for doubt that suspect agent is capable of impairing the structure or function of the nervous system. This is a rather severe statement as morphological changes are required. With these criteria, only a small number of solvents can be regarded as neurotoxic. The following table is taken from the paper of Spencer and Schaumburg:

Neurotoxic effects of different solvents
(H = human, A = animal)

Proven Human Neurotoxicants:
carbon disulphide: psychosis (H), neuropathy (H,A), tardive parkinsonism (H)
n-hexane + methyl-ethyl ketone: peripheral neuropathy (H, A)
methyl-n-butyl ketone + methyl-ethyl ketone: peripheral neuropapthy (H, A)
toluene (substance abuse only): auditory (A), brainstem, cerebellar, pyramidal and other types of irreversible brain dysfunction (H)
impure trichloroethylene: trigeminal neuralgia (H), transverse myelopathy (H?)

Proposed Human Neurotoxicants
diethyl ether: anorexia ? (H)

<u>ethylene chloride:</u> encephalopathy, tremor ? (H)
<u>nitrobenzene:</u> limb numbness (H)
<u>pyridine:</u> facial paralysis ? (H)
<u>styrene:</u> ototoxicity (A), neuropathy (H)
<u>tetrachloroethane:</u> neuropathy ? (H)
<u>trichloroethylene + perchloroethylene:</u> organic mental syndrome ? (H)
<u>xylene:</u> ototoxicity (A), organic mental syndrome ? (H)
<u>white spirit:</u> organic mental syndrome (H)
<u>various mixed solvents:</u> organic mental syndrome ? (H)

According to this table, only carbon disulphide, toluene and trichloroethylene have proven toxic effects upon the central nervous system. The list of proposed neurotoxicants contains a number of solvents probably causing "organic mental syndrome", which is the same as the neuroasthenic syndrome.

Effects of long-term exposure to solvents on the central nervous system in man - epidemiological studies

When we (Elofsson et al 1980) studied a group of spray painters we found a rather wide-spread involvement of the nervous system: Effects on performance in behavioral tests measuring perceptual speed, memory, manual dexterity and reaction time, on psychiatric status indicating presence of neurasthenic syndrome, on visual evoked potentials, on EEG, on computed tomography. Almost all physiological measurements and clinical tests or examinations showed influence upon the central nervous system in the exposed group as compared to a matched control group. The exposure in this study was a mixture of between 8 and 12 different solvents. The solvents most commonly found were the following (average concentration given as a proportion of the hygienic limit value): Methyl ethyl ketone (0.25), white spirit (0.17), toluene (0.13), methylene chloride (0.11), methyl ethyl ketone (0.06), trichloroethylene (0.05), xylene (0.04), styrene (0.04), butanol (0.04), ethanol (0.01), ethyl acetate (0.01), butyl acetate (0.01). The total amount of solvents was in average 0.30, i.e. not more than one third of the Swedish TLVs. Just a few solvents were present in all samples: toluene, xylene and (for the car painters) ethyl acetate.

In another study (Wennberg et al 1983) we examined in a similar way rotogravure printers, who were almost exclusively exposed to toluene. The exposure level was around the Swedish TLV, which means around three times as much solvents and ten times as much toluene as the abovementioned spray painters. Nevertheless, the demonstrated effects upon the central nervous system were much more limited as compared to those of the painters. Thus were there no effects found upon the psychiatric status, upon the visual evoked responses, upon the EEG or upon the performance in a majority of the tests. Only one single test showed an impairment compared to the controls: simple reaction time, that was prolonged - even more than in the painter group.

At the international conference on organic solvent toxicity in Stockholm 15-17 October 1984, several similar studies were presented and most of them indicate the presence of effects upon the central nervous system:

Konietzko et al from Mainz in West-Germany found effects on visual evoked potentials in a group of 36 floorers.

Örbaek et al from Malmö/Lund in Sweden found several signs of effects among 50 workers exposed to a mixture of solvents and the conclude "that organic brain effect on long term exposure to organic solvents are clearly demonstrated"

Hänninen from Helsinki in Finland found relations between exposure to solvents and impairment of performance in psychomotor tests in a group of 132 factory workers.

Mikkelsen et al from Copenhagen in Denmark found the prevalence of organic psychosyndrome three times higher in a group 85 of painters than in a similar group of bricklayers.

Rasmussen et al from Aarhus in Denmark examined a group of 240 metal workers exposed to chlorinated solvents. For all four solvents (trichlorethylene, 1,1,1-trichlorethane, methylene chloride and freon) they found significantly higher frequency of symptoms of acute and chronic neuropsychological effects compared to a control group.

Antti-Poika et al from Helsinki in Finland examined a group of 43 photogravure printers exposed to toluene. They found no significant effects of this exposure. They performed clinical neurological examination, EEG, evoked potentials, psychological tests and computed tomography

Silberschmid et al from Slagelse in Denmark examined 39 female printers exposed to cyclohexanone, xylene and toluene at levels just below their TLVs. They found signs of psycho-organic syndrome (i.e. neuro-asthenic syndrome).

Despite the fact that many independent studies now clearly indicates that long term exposure to solvents at levels at or below actual TLVs may cause disturbances in the functions of the central nervous system, still some questions concerning this remain to be solved:

(1) Can all solvents cause the same type of damage or are there specific target points in the brain for different solvents?

(2) Which are the basic physiologic mechanisms in solvent toxicity on the central nervous system?

(3) Are there synergistic effects between different solvens or will a mixture of solvents cause an effect amounting to the sum of the effects of the separate solvents in the mixture?

(4) Are the toxic effects of long term low level exposure to solvents chronic or reversible?

(5) Is it possible to put the diagnosis chronic solvent intoxication (or encephalopathy) in a clinical case. To-day this is a probability diagnosis.

The future research in the field of solvent toxicity has to cover these questions. This can be done in further epidemiological studies but also, which might be the major part of the future work, by animal experiments. Several such experimental projects are already going on.

Experimental studies on toxicological mechanisms by solvents in the central nervous system

Our knowledge about the basic mechanisms of solvent toxicity in the central nervous system is presently very limited. However, knowledge of basic mechanisms is important in understanding impairments of higher brain functions in man. A majority of the experimental studies are performed on animals. In Sweden this type of research has been running or recently started at several research centra. The Swedish Work Environment Fund has recently decided to give priority to toxicological research by initiating research programs within this field and special priority has been given to three branches of toxicology, neurotoxicology being one of those. This will, of course, turn over a considerable part of the total research power in Sweden to problems concerning basic mechanisms of solvent neurotoxicity.

There are presently several groups in Sweden focussed on solvent neurotoxicity, the following being some of the most important:

(1) At the National Board of Occupational Safety and Healt (NBOSH) and the Karolinska Institute, Bengt Knave and Urban Ungerstedt with co-workers have tried to mimic the real occupational situation by exposing rats during long time (almost life-time) for fairly low levels of pure solvents or a mixture of two different solvents. Urban Ungerstedt has developed a very delicate technique of microdialysis of the brain, by which it i possible to detect and estimate small amounts of i.e. neurotransmitters from stereotactically identified areas in the brain of a living rat. These results can then be related to simultaneously performed behavioral and neurophysiological measurements.

(2) Fuxe at the Karolinska Institute and his collaborators have shown that benzene, ethylbenzene, xylene (orto, meta and para) can produce selective changes in certain types of cathecolamine neuron systems and that toluene in low concentrations can reduce dopamine turnover in certain striatal dopamine nerve terminal systems as well as increase noradrenaline turnover in hypothalamic nerve terminal systems.

(3) Haglid and his group at Göteborg University have studied effects of solvents upon CNS and they have shown that exposure to trichloroethylene, perchloroethylene, methylene chloride as well as xylene and toluene at moderate exposure levels results in irreversible astroglial hypertrophy and/or proliferation. They have also demonstrated effects by solvents upon phospholipids that constitutes one of the components in the inner surface of the cell membrane.

(4) At Lund University, Alling and co-workers have focussed their interest upon solvent effects upon neuropeptides. These could bind to membrane lipids such as cerebrosides, gangliosides and phospholipids and alter their configuration and that of adjoining proteins, an action that affects functional characteristics of neurotransmitter receptors and other synapse and membrane components. No results hade so far been published but the approach is interesting and important concerning the future understanding of solvent toxicology.

Central or peripheral effects?

Sometimes it might be difficult to decide whether an effect upon the nervous system is caused primarily by an action upon the central nervous system or peripheral nerves. We studied the effect upon the sense of smell in tank cleaners exposed to fumes of heated heating oil and found a defect in the sense of smell similar to the loudness recruitment in hearing loss, i.e. normal perception at strong stimuli but impaired perception at weak stimuli. Is this caused by an effect upon the sensory cells in the olfactory mucosa or is it caused by solvent upon the brain via uptake in the lungs and the blood? Is strong smelling solvents more hazardous to the sense of smell than weak smelling?

Literature References

Elofsson S-A, Gamberale F, Hindmarsh T, Iregren A, Isaksson A, Johnsson I, Knave B, Lydahl E, Mindus P, Persson H E, Philipson B, Steby M, Struwe G, Söderman E, Wennberg A, Widén L: Exposure to organic solvents - a cross-sectional epidemiologic investigation on occupationally exposed car and industrial spray painters with special reference to the nervous system. Scand j work environ health 6 (1980) 239-273.

ICOST, International Conference on Organic Solvent Toxicity 15-17 Oct 1984, Stockholm, Sweden.

Spencer P S and Schaumburg H H: Organic solvent neurotoxicity: Facts and research need. Invited paper at ICOST. - Scand j work environ health 11 Suppl 1 (in press).

Wennberg A et al: Exposure to a mixture of solvents and to a single solvent - a comparison of neurotoxic effects on industrial workers. Poster presented att ICOST.

Safety and Health Aspects of Organic Solvents, pages 247–253

SOLVENTS AND PERIPHERAL NEUROPATHY

Anna Maria Seppäläinen

Division of Clinical Neurophysiology
Department of Neurology, University of Helsinki,
SF-00290 Helsinki, Finland

The peripheral nervous system (PNS) is a less complicated object for study than the central nervous system (CNS). The clinical picture of polyneuropathy is often clear, although at times it is difficult to differentiate sensory symptoms of the PNS from those of the CNS on clinical grounds only. The extent and localization of neuropathy can be determined rather quickly and reliably with electroneuromyography, that is with measurement of motor and sensory nerve conduction velocities and by studying electromyography (EMG) of a few muscles. In clinical cases of polyneuropathy nerve biopsy of the sural nerve can be taken and analyzed. Neurophysiological, neuropathological and neurochemical methods can also be applied to animal studies, where experimental exposure may cause peripheral neuropathy. In experimental studies we have more precise knowledge about the dose and can use pure chemical compounds. Experimental studies can be applied to clarify toxic mechanisms and to ascertain etiological connections between certain solvents and neuropathy. Epidemiological studies among occupationally exposed workers and relevantly chosen referents also clarify etiological connections and give information of human effects of solvents. Deduction from animal studies, where high dosages and short exposures are often used, to effects of a long-term, low-dose human exposure in an occupational setting is not simple and may result in wrong conclusions.

The clearest evidence of solvent neuropathy has been collected on hexacarbons (n-hexane, methyl-n-butyl ketone and their metabolites) and carbon disulfide (Billmaier et al., 1974; Cianchetti et al. 1976; Herskowitz et al., 1971; Spencer et al., 1980; Seppäläinen and Haltia, 1980). It is also known that methyl-ethyl ketone potentiates the neurotoxic effect of hexacarbons (Altenkirch et al., 1977). These solvents cause clinical neuropathy with weakness of distal muscles and paresthesiae and numbness especially in glove-stocking distribution. In mild cases only occasional numbness of hands and feet is complained of. Sensory conduction velocity (SCV) is diminished especially in the distal nerves, and the amplitude of the sensory action potential is also diminished. Motor conduction velocity (MCV) may as well be reduced, first in the longest axons, i.e., in leg nerves such as the deep peroneal nerve (Allen et al., 1975; Cianchetti et al., 1976). Electromyography reveals denervation activity (fibrillations and positive sharp waves), and with long standing neuropathy the number of acting motor units is reduced, and the duration of remaining motor units increases. Neuropathological studies have revealed central peripheral distal axonopathy signified by the accumulation of 10 nm neurofilaments close to the terminal portion of the axons (Spencer et al., 1980). This accumulation of neurofilaments above the crushed region was probably the lesion responsible for slower return of function and severely reduced number of regenerating myelinated fibers during the first 4 weeks of regeneration in rats chronically poisoned with 2,5-hexanedione in connection with crushing of the posterior tibial nerve (Simonati et al., 1983). Both the electrophysiological and neuropathological findings have been similar in neuropathy caused by hexacarbons and by carbon disulfide (Seppäläinen and Haltia, 1980).

Several epidemiological studies have demonstrated that workers with long-term exposure to carbon disulfide have slowed nerve conduction velocities, electromyographic abnormalities as well as complaints of paresthesiae, diminished strength and pain in the limbs more often than their referents (for review see Seppäläinen and Haltia, 1980). Mild neuropathic findings have already been detected when exposure levels have been around 20 - 40 ppm (parts per million), mostly these workers had been active in working life at the time of the studies. Neuropathic findings have, however, been quite stable since they were still shown among workers whose exposure to carbon disulfide had ceased 10 years earlier (Corsi et al., 1983).

Central peripheral distal axonopathy is not limited to the PNS. Probably similar lesions in the visual pathways were the basis for abnormal visual evoked potentials (VEP) among workers with repeated short periods of occupational exposure to concentrations of up to 2000 - 3000 ppm of n-hexane (Seppäläinen et al., 1979). Experimental studies have shown degeneration in the visual tracts of cats poisoned with hexacarbons (Schaumburg and Spencer, 1978). Also the early components of somatosensory evoked potentials (SEP) were slightly delayed and the later components of SEPs flatter among workers exposed to solvent fumes containing n-hexane in excessive amounts (Mutti et al., 1982). Motor and sensory conduction velocities were also significantly slower among the exposed in comparison to age-matched referents in this study, but all the SEP changes could not be explained by peripheral neuropathy. This also suggests a related CNS effect of hexacarbons in addition to the PNS effects.

Among patients with solvent poisoning symptoms and signs of peripheral neuropathy have been found, although the solvent mixtures used have not contained relevant amounts of hexacarbons or carbon disulfide (Seppäläinen et al., 1980; Seppäläinen and Antti-Poika, 1983; Feldman et al., 1979). Most of these patients have been exposed to mixtures of solvents containing usually toluene, xylene, various aliphatic hydrocarbons and at times halogenated hydrocarbons.

A follow-up study of patients with solvent poisoning revealed that at the time of diagnosis a very high proportion of the patients had neuropathic signs - slowed nerve conduction velocities and/or neurogenic abnormalities in their electromyography, and that these types of abnormalities were of long standing (Seppäläinen and Antti-Poika, 1983). Actually, a slightly higher proportion of the patients showed some neuropathic signs 6 years later (74% versus 62% in the first examination), even though the occupational solvent exposure of most patients had ceased within 1-2 years of the diagnosis. Some improvement of the findings was noted for 43% of the patients, whereas deterioration was found for 33% of them. Usually the deterioration was only mild. Thus neuropathic signs were quite permanent after exposure to various mixtures of solvents as it was after occupational exposure to carbon disulfide. A small group of patients, altogether 13 of them had been exposed both to mixtures of solvents and to chlorinated hydrocarbons. Their neuropathy seemed more extensive, it was seen also in the proximal segments and it involved always also motor fibers when compared to other patients in the study, who had been exposed either to solvent mixtures or to solely chlorinated hydrocarbons. The number of patients in subgroups was rather small, consequently no definite conclusions can be made. The finding, however, suggested a potentiating effect of solvent mixtures in combination with chlorinated hydrocarbons. The toxic interactions that arise after mixed exposure are poorly known, but they may play an important role in producing clinical symptoms and signs.

In epidemiologic studies conducted among groups of workers occupationally exposed to solvent mixtures, e.g. car painters, industrial spray painters, some individuals have been detected who have shown abnormally slow conduction velocities in one or more nerves, or the mean conduction velocity of the exposed group has been statistically significantly slower than that of the reference group. These types of findings were detected in Swedish (Elofsson et al., 1980) and Finnish (Seppäläinen et al., 1978) studies in which the time weighted average exposure level to solvents was about 30% of the hygienic standard.

Italian studies have shown certain exposure-response relationships between solvent exposure and neuropathic findings; the latter increased when the amount of glue/worker/day in shoe industry was greater than 1.3 kg and ventilation was poor (Buiatti et al., 1978). Among patients with solvent poisoning the frequency of abnormal nerves increased as the exposure levels were higher (Seppäläinen et al., 1980). Among repair painters in construction industry no excess of polyneuropathy was detected in comparison to concrete workers with no solvent exposure (Seppäläinen and Lindström, 1982), however, during recent years the exposure to organic solvents has been rather low in house painting as mainly latex paints have been used.

The exposure-response relationships in solvent exposure are not easily revealed. Often the individual exposure intensity cannot be determined, the overall exposure levels may vary greatly during the working days and the individual working habits of the worker may affect his individual exposure. In occupational exposure the effects of occasional or frequent peak exposure periods cannot be reliably separated. It is probable that repeated short periods of high exposure may cause more damage than stable low-level exposure. As the development of neuropathy is usually slow and signs appear only after several months or years of occupational exposure, we should be able to evaluate reliably also the past exposures to clarify exposure-effect relationships in long-term exposure.

At present exposure levels to solvents are often low in modern industry, but even in those settings slight effects in the peripheral nerves of some workers can be detected with sensitive functional methods. That may partly depend on varying individual sensitivity towards chemicals. In small work places where ventilation may not be effective enough or where information about chemical hazards has not reached people, clinical signs of neuropathy may still develop. As the regeneration capacity of the nervous system is limited we should be aware of those hazards and use sensitive neurophysiological methods to detect even slight stages of disease. With early detection extensive damage can be prevented and also the amelioration of the symptoms and signs is more probable.

REFERENCES

Altenkirch H, Mager J, Stoltenburg G, Helmbrecht J (1977). Toxic polyneuropathies after sniffing a glue thinner. J Neurol 214: 137-152.

Billmaier D, Yee HT, Allen N, Craft B, Williams N, Epstein S, Fontaine R (1974). Peripheral neuropathy in a coated fabrics plant. J Occup Med 16: 665-671.

Buiatti E, Cecchini O, Ronchi O, Dolara P, Bulgarelli G (1978). Relationship between clinical and electromyographic findings and exposure to solvents in shoe and leather workers. Br J Ind Med 35: 168-173.

Cianchetti C, Abbritti G, Perticoni A, Siracusa A, Curradi F (1976). Toxic polyneuropathy of shoe-industry workers: A study of 122 cases. J Neurol Neurosurg Psychiatry 39: 1151-1161.

Corsi G, Maestrelli P, Picotti G, Manzoni S, Negrin P (1983). Chronic peripheral neuropathy in workers with previous exposure to carbon disulphide. Br J Ind Med 40: 209-211.

Elofsson SA, Gamberale F, Hindmarsch T, Iregren A, Isaksson A, Johnsson I, Knave B, Lydahl E, Mindus P, Persson HE, Philipson B, Steby M, Struwe G, Söderman E, Wennberg A, Widen L (1980). Exposure to organic solvents. A cross-sectional epidemiologic investigation on occupationally exposed car and industrial spray painters with special reference to the nervous system. Scand J Work Environ Health 6: 239-273.

Feldman RG, Mayer RM, Taub A (1970). Evidence for peripheral neurotoxic effect of trichloroethylene. Neurology 20: 599-606.

Herskowitz A, Ishii N, Schaumburg H (1971). n-Hexane neuropathy. A syndrome occurring as a result of industrial exposure. New Engl J Med 285: 82-85.

Mutti A, Ferri F, Lommi G, Lotta S, Lucertini S, Franchini I (1982). n-Hexane-induced changes in nerve conduction velocities and somatosensory evoked potentials. Int Arch Occup Environ Health 51: 45-54.

Schaumburg HH, Spencer PS (1978). Environmental hydrocarbons produce degeneration in cat hypothalamus and optic tract. Science 199: 199-200.

Seppäläinen AM, Antti-Poika M (1983). Time course of electrophysiological findings for patients with solvent poisoning: A descriptive study. Scand J Work Environ Health 9: 15-24.

Seppäläinen AM, Haltia M (1980). Carbon disulfide. In Spencer PS, Schaumburg HH (eds): "Experimental and Clinical Neurotoxicology," Baltimore: Williams & Wilkins, pp 356-373.

Seppäläinen AM, Husman K, Mårtenson C (1978). Neurophysiological effects of long-term exposure to a mixture of organic solvents. Scand J Work Environ Health 4: 304-314.

Seppäläinen AM, Lindström K (1982). Neurophysiological findings among house painters exposed to solvents. Scand J Work Environ Health 8: suppl. 1, 131-135.

Seppäläinen AM, Lindström K, Martelin (1980). Neurophysiological and psychological picture of solvent poisoning. Am J Industr Med 1: 31-42.

Seppäläinen AM, Raitta C, Huuskonen MS (1979). n-Hexane-induced changes in visual evoked potentials and electroretinograms of industrial workers. Electroenceph Clin Neurophysiol 47: 492-498.

Simonati A, Rizzuto N, Cavanagh JB (1983). The effects of 2,5-hexanedione on axonal regeneration after nerve crush in the rat. Acta Neuropath (Berl) 59: 216-224.

Spencer PS, Couri D, Schaumburg HH (1980). n-Hexane and methyl n-butyl ketone. In Spencer PS, Schaumburg HH (eds): "Experimental and Clinical Neurotoxicology," Baltimore: Williams & Wilkins, pp 456-475.

Safety and Health Aspects of Organic Solvents, pages 255–264

SYMPTOMS AND SIGNS IN SOLVENT EXPOSED POPULATIONS

Mari Antti-Poika

Institute of Occupational Health, Haartmanink.1,
SF-00290 Helsinki, Finland

This paper is restricted only to the symptoms and signs observed in the context of a long-term occupational exposure. Thus, acute effects are not considered neither the effects among solvent sniffers which are diversified and caused by exposure levels much in excess of those normally occuring in the workplace. The reported symptoms and signs in a long-term solvent exposure are mainly neurological and neuropsychiatric.

The best known manifestations of solvent toxicity are those caused by carbon disulfide, n-hexane, and methyl n-butyl ketone. Carbon disulfide is known to cause multifocal nervous system damage with both central and peripheral nervous system effects (Vigliani, 1954). In addition, it has been shown to be associated with an increased prevalence of coronary heart disease (Tolonen et al., 1979). In the cases of n-hexane and methyl n-butyl ketone intoxication, the most prominent finding has been the peripheral polyneuropathy but, in some cases, also signs of central effects have been found (Allen et al., 1975, Cianchetti et al., 1976, Sobue et al., 1978, Passero et al., 1983).

A more problematic question is the association of mild unspecific symptoms and signs with exposure to various solvent mixtures. Chronic solvent effects have been diagnosed as occupationed diseases in the Nordic countries (see, Juntunen, in this volume) but the concept of "chronic organic solvent intoxication" has been challenged elsewhere. The exposures, involved in both case studies and epidemiological studies, have varied. Most commonly, they have constituted mixtures of aliphatic hydrocarbons with variable amounts of aromatic hydrocarbons, and intoxications caused by aromatic hydrocarbons as well as chlorinated hydrocarbons have also been diagnosed. No specific effects of any of the agents can be isolated.

CASES WITH DIAGNOSED CHRONIC INTOXICATION

It should be emphasized that based on case reports, no etiological conclusions can be drawn. In all of these studies, the patients were selected. At least some symptoms and signs have contributed to reaching the diagnosis and thus it is natural that these findings are overrepresented in the studied group.

In most studies, common symptoms have been related to the central nervous system: disturbances in memory and concentration, abnormal fatigue, changes in mood, and headache (Table 1). On the other hand, the clinical signs

TABLE 1. Symptoms in patients with diagnosed chronic organic solvent intoxication.

Disturbances in memory and concentration, abnormal fatigue, depression, irritability, psychic lability, headache	Gregersen et al. 1978 (35) Arlien - Soborg et al. 1979 (50) Juntunen et. al 1980 a (37) Antti-Poika 1982 a (87)
Sensory disturbances	Juntunen et al. 1980 a (37) Antti-Poika 1982 a (87)
Nausea, tremor	Antti-Poika 1982 a (87)

(Number of patients shown in parenthesis)

have varied (Table 2), probably mainly according to which aspects have been emphasized in the examinations. In the Danish studies of Arlien-Soborg and others, tremor, polyneuropathy and vestibular dysfunction have been reported. In Finnish patients, the so-called psycho-organic syndrome and cerebellar dysfunction have been the most prominent signs.

TABLE 2. Clinical signs in patients with diagnosed chronic organic solvent intoxication.

Tremor Polyneuropathy Vestibular dysfunction	Arlien - Soborg et al. 1979 (50)
Vestibular dysfunction	Arlien - Soborg et al. 1981 (113)
Slight psycho-organic alteration Cerebellar dysfunction Peripheral neuropathy Cranial nerve disturbances	Juntunen et al. 1980 a (37)
Psycho-organic syndrome Cerebellar dysfunction Disturbances of the fine motor function Disturbances in gait and station	Antti-Poika 1982 b (87) Juntunen et al. 1982 (80)

(Number of patients examined shown in parenthesis)

In some patients, radiological signs of brain atrophy have also been detected (Table 3). We should remember, however, that the finding of brain atrophy has, in all the cases, indicated the diagnosis of solvent intoxication which makes it impossible to draw any etiological conclusions. Gade et al. found in their controlled study that the picture of intellectual impairment and brain atrophy is by no means specific to solvent exposure.

TABLE 3. "Brain atrophy" in patients with diagnosed chronic organic solvent intoxication.

Examined group	Finding	Reference
18 patients exposed to solvent mixtures (12 PEG, 6 CT)	Central and/or cortical atrophy in 17	Gregersen et al. 1978
50 patients exposed to solvent mixtures (12 PEG, 38 CT)	Central and/or cortical atrophy in 31	Arlien-Soborg et al. 1979
37 patients exposed mainly to solvent mixtures (37 PEG)	Central and/or cortical atrophy in 23 No dose - response	Juntunen et al. 1980b
31 patients with brain atrophy and diagnosed "toxic encephalopathy" due to solvents, + 31 patients with brain atrophy without solvent exposure, + 31 normal controls	No difference in intellectual impairment between the groups with atrophy	Gade et al. 1984

The terminology of clinical diagnostics and their meanings have caused considerable confusion in the Nordic countries and elsewhere. The Danish investigators have used the term "dementia." The Swedish and Finnish researchers have used "psycho-organic syndrome" but they mean slightly different matters. In Finland, the concept of psycho-organic syndrome is primarily a clinical neurological entity and confirmed by psychological tests; in Sweden the corresponding criteria encompass positive symptoms and pathological results in psychological tests. The Danish term dementia means intellectual impairment in psychological tests.

EXPOSED AND NONEXPOSED POPULATIONS IN EPIDEMIOLOGICAL STUDIES

Table 4 presents the results of some studies where exposed groups have been compared with nonexposed groups. In most of the studies, symptoms have been more prevalent in the exposed groups. Psychological performances have also usually been poorer in the exposed groups. The results of the neurophysiological studies have varied. In two studies computerized tomography (CT) of the brain has been used and in both of them the control group had wider sulci than the exposed group. Gregersen et al. found

TABLE 4. Differences between the exposed and unexposed groups in separate examinations.

Exposure (Number of exam.)	Symptoms	Signs	Psych	EEG	ENMG	CT	Other	References
Styrene (98+98)	+		+	+	-			Seppäläinen et al. 1976 Lindström et al. 1976 Härkönen 1977
Mixtures (102+102)	+	+	+	?	+			Hänninen et al. 1976 Seppäläinen et al. 1978 Husman 1980 Husman, Karli 1980
Jet fuel (30+30)	+	-	+	+	±			Knave et al. 1978
Mixtures (80+80+80)	+	?	+	±	+	-		Elofsson et al. 1980
Mixtures (219+229)	-		+	-	-			Lindström & Seppäläinen 1980
Mixtures (250+200)	+							Hane, Hogstedt 1980 Hogstedt et al. 1980
Mixtures (42+42)	+							Agrell et al. 1980
Organic solvents (65+33)	+	-	+				CA+	Gregersen et al. 1984
Toluene (43+31)	±	-	+	-		-	ANS-	Antti-Poika et al. 1985 Juntunen et al. 1985
Mainly mixtures (34+20)							ANS+	Matikainen & Juntunen 1985

CA = Cerebral asthenopia
ANS = Autonomic nervous system dysfunction

increased cerebral asthenopia (changes in the visual experience of a fixed object after a certain period of time) in the exposed group. Small groups have been examined for the autonomic nervous system (ANS). Matikainen et al. found disturbances in the ANS in workers exposed mainly to solvent mixtures and especially in persons who also had signs of peripheral polyneuropathy. Disturbances in the ANS were not found in workers exposed to toluene.

The symptoms, more prevalent in the exposed groups, have been rather similar in the separate studies and also similar to the symptoms of the patients with solvent intoxication (Table 5). Again, it must be noted that

TABLE 5. Symptoms occuring differently in the exposed and unexposed groups.

Exposure	Symptoms	Reference
Styrene	Fatigue, difficulties in concentration, symptoms of irritation	Härkönen 1977
Mixtures	Fatigue, disturbances in memory and concentration	Husman 1980
Jet fuel	Fatigue, depression, loss of drive, palpitations	Knave et al. 1978
Mixtures	Fatigue, worrying, difficulties in memory and concentration, headache, nausea, epigastric pain	Elofsson et al. 1980
Mixtures	Fatigue, depression, irritability, memory disturbances, unclear chest pains, diminished sexual activity	Hane, Hogstedt 1980, Hogstedt et al. 1980 Agrell et al. 1980
Org. solvents	Fatigue, disturbances in memory, irritability, emotional lability	Gregersen et al. 1984

the very same symptoms have been looked for which likely increases the uniformity of the results. The clinical signs which differed between the exposed and nonexposed groups in one study were psycho-organic syndrome and disturbances in sensibility (Husman & Karli 1980).

There is evidence of an increased number of neurological and neuropsychiatric symptoms and signs in solvent exposed populations. A totally separate question is, however, the diagnosis of individual cases of solvent intoxication, which is described elsewhere in this issue (Juntunen). Early diagnosis of slight cases of organic solvent intoxication as accurately as possible is a great challenge to occupational medicine.

REFERENCES

Agrell A, Hane M, Hogstedt C (1980). Symptoms among housepainters - a five year follow-up study. Läkartidningen 77:440-442 (in Swedish, English summary p. 442).

Allen M, Mendell JR, Billmaier DJ, Fontaine RE, O'Neill J (1975). Toxic polyneuropathy due to methylbutyl ketone. Arch Neurol 32:209-222.

Antti-Poika M (1982a). Prognosis of symptoms in patients with diagnosed chronic organic solvent intoxication. Int Arch Occup Environ Health 51:81-89.

Antti-Poika M (1982b). Overall prognosis of patients with diagnosed chronic organic solvent intoxication. Int Arch Occup Environ Health 51:127-138.

Antti-Poika M, Juntunen J, Matikainen E, Suoranta H, Hänninen H, Seppäläinen AM, Liira J. Occupational exposure to toluene-neurotoxic effects with special emphasis on drinking habits. Int Arch Occup Environ Health (submitted for publication).

Arlien-Soborg P, Bruhn P, Gyldenstedt C, Melgaard B (1979). Chronic painter's syndrome. Acta Neurol Scand 60:149-156.

Arlien-Soborg P, Zilstorff K, Grandjean B, Milling Pedersen L (1981). Vestibular dysfunction in occupational chronic solvent intoxication. Clin Otolaryngol 6:285-290.

Cianchetti C, Abbritti G, Perticoni G, Siracusa A, Curradi F (1976). Toxic polyneuropathy of shoe-industry workers. A study of 122 cases. J Neurol Neurosurg Psychiatry 39:1151-1161.

Elofsson S-A, Gamberale F, Hindmarsh T, Iregren A, Isaksson A, Johnsson I, Knave B, Lydahl E, Mindus P, Persson HE, Philipson B, Steby M, Struwe G, Söderman E, Wennberg A, Widén L (1980). Exposure to organic solvents. A cross-sectional epidemiologic investigation on occupationally exposed car and industrial spray painters with special reference to the nervous system. Scand J Work Environ Health 6:239-273.

Flodin U, Edling C, Axelson O (1984). Clinical studies of psychoorganic syndromes among workers with exposure to solvents. Am J Ind Med 5:287-295.

Gade A, Lykke Mortensen E, Browne E, Udesen H (1984). The pattern of intellectual impairment in toxic encephalopathy. An abstract. International congress on organic solvent toxicity. Arbete och Hälsa 29, 95.

Gregersen P, Angelso B, Elmo Nielsen T, Norgaard B, Uldal C (1984). Neurotoxic effects of organic solvents in exposed workers. An occupational, neuropsychological, and neurological investigation. Amer J Ind Med 5:201-225.

Gregersen P, Mikkelsen S, Klausen H, Dossing M, Nielsen H, Thygesen P (1978). Et kronisk cerebralt malersyndrom. Ugeskr. Laeg 140:1638-1644 (in Danish, English summary p. 1644).

Hane M, Hogstedt C (1980). Subjective symptoms among occupational groups exposed to organic solvents. Läkartidningen 77:435-436 (in Swedish, English summary p. 439, references pp. 441-442).

Hänninen H, Eskelinen L, Husman K, Nurminen M (1976). Behavioral effects of long-term exposure to a mixture of organic solvents. Scand J Work Environ Health 2:240-255.

Härkönen H (1977). Relationship of symptoms to occupational styrene exposure and to the findings of electroencephalographic and psychological examinations. Int Arch Occup Environ Health 40:231-239.

Hogstedt C, Hane M, Axelson O 1980. Diagnostic and health care aspects of workers exposed to solvents. In Zenz (ed): "Developments in Occupational Medicine." London: Year Book Medical Publishers, pp. 249-258.

Husman K (1980). Symptoms of car painters with long-term exposure to a mixture of organic solvents. Scand J Work Environ Health 6:19-32.

Husman K, Karli P (1980). Clinical neurological findings among car painters exposed to a mixture of organic solvents. Scand J Work Environ Health 6:33-39.

Juntunen J, Antti-Poika M, Tola S, Partanen T (1982). Clinical prognosis of patients with diagnosed chronic solvent intoxication. Acta Neurol Scand 65:488-503.

Juntunen J, Hernberg S, Eistola P, Hupli V (1980b). Exposure to industrial solvents and brain atrophy. Europ Neurol 19:366-375.

Juntunen J, Hupli V, Hernberg S, Luisto M (1980a). Neurological picture of organic solvent poisoning in industry. Int Arch Occup Environ Health 46:219-231.

Juntunen J, Matikainen E, Antti-Poika M, Suoranta H, Valle M. Nervous system effects of long-term occupational exposure to toluene. Acta Neurol Scand (submitted for publication).

Knave B, Olson BA, Elofsson S, Gamberale F, Isaksson A, Mindus P, Persson HE, Struwe G, Wennberg A, Westerholm P (1978). Long-term exposure to jet fuel. II. A cross-sectional epidemiologic investigation on occupationally exposed industrial workers with special reference to the nervous system. Scand J Work Environ Health 4:19-45.

Lindström K, Härkönen H, Hernberg S (1976). Disturbances in psychological functions of workers occupationally exposed to styrene. Scand J Work Environ Health 2:129-139.

Lindström K, Seppäläinen AM (1980). Study of concrete reinforcement workers and maintenance house painters. Part 4: Symptoms, psychological functions and neurophysiological findings among maintenance house painters exposed to solvents. Investigation of Institute of Occupational Health no 170, Helsinki, 59 p. (in Finnish, English summary pp. 53-55).

Matikainen E, Juntunen J. Autonomic nervous system dysfunction in workers exposed to organic solvents. (A manuscript).

Passero S, Battistini N. Cioni R, Giannini F, Paradiso C, Battista F, Carboncini F, Sartorelli E (1983). Toxic polyneuropathy of shoe-industry workers in Italy. Italian J Neurol Sci 4:463-472.

Seppäläinen AM, Härkönen H (1976). Neurophysiological findings among workers occupationally exposed to styrene. Scand J Work Environ Health 2:140-146.

Seppäläinen AM, Husman K, Mårtenson C (1978). Neurophysiological effects of long-term exposure to a mixture of organic solvents. Scand J Work Environ Health 4:304-314.

Sobue I, Iida M, Yamamura Y, Takayanagui T (1978). N-hexane polyneuropathy. Int J Neurol 11:317-330.

Tolonen M. Nurminen M, Hernberg S (1979). Ten-year coronary mortality of workers exposed to carbon disulfide. Scand J Work Environ Health 5:109-114.

Vigliani EC (1954). Carbon disulphide poisoning in viscose rayon factories. Brit J Ind Med 11:235-244.

Safety and Health Aspects of Organic Solvents, pages 265–279

OCCUPATIONAL SOLVENT POISONING: CLINICAL ASPECTS

Juhani Juntunen

Clinical Neurosciences, Institute of Occupational Health, SF-00290 Helsinki, Finland

INTRODUCTION

The extensive application of new organic solvents and their mixtures in industrial communities each year has given rise to many previously unsuspected toxic effects. The acute narcotic effects of solvents and some outbreaks of specific neurological diseases among particular groups of exposed industrial workers are well known. Particularly reports from the Nordic countries have suggested that long-term occupational exposure to organic solvents is associated with nonspecific neurologic and psychiatric symptoms among the workers. In many countries, a diffuse neuropsychiatric disease associated with long-term occupational exposure to solvents is generally accepted as occupational disease, while the diagnostic criteria used seem to vary greatly from country to country. This has created much controversy and even debate during the past few years among investigators in this field.

In this article, a short review of the occupational solvent poisoning is given. The main emphasis is on the clinical characteristics of chronic solvent poisoning at an individual level.

GENERAL ASPECTS OF NEUROTOXIC SOLVENTS AND THEIR EFFECTS ON THE NERVOUS SYSTEM

Organic solvents constitute a large group of volatile chemicals, mainly hydrocarbons and their derivates. They are characterized by their nonpolarity and lipid solubility. The usual road of entry is by inhalation, but some solvents may also penetrate through the skin. The nature of work tasks, the state of cardiovascular and respiratory systems, and individual factors determine the rate at which organic solvents are metabolized (e.g. Fiserova-Bergerova, 1985). Organic solvents can reach the nervous system by penetrating the blood-neural barrier which protects most sites of the central nervous system and the peripheral nervous system (excluding dorsal root ganglia and autonomic ganglia). There are only few solvents which have been definitely shown to cause peripheral neuropathy in occupational settings (see later). When the central nervous system toxicity of solvents is considered, it should be emphasized that agents which can penetrate brain tissue do no necessarily equally affect all cell types in the brain, as various areas have different susceptibility.

In work environments, most organic solvents are used as mixtures: lacquer thinners, petroleum ether, white spirit, mineral terpentine, and greasing oils. These mixtures generally contain various solvents in many combinations, mostly toluene, methyl-isobutyl ketone, isobutanol, acetone, ethylene glycol, isopropanol, xylene, butylacetate etc. Consequently, great difficulties arise always when the individual exposure is assessed. In this assessment, a careful consideration of the chemistry of the compounds used, the nature of work tasks and methods, the work conditions, the protective equipment used, and the results of ambient air measurements and biological monitoring should be performed. All these data should be weighed against previous experience with similar exposures (e.g. Järvisalo and Tossavainen, 1982). Table 1 shows the most important solvents that cause problems in occupational neurology and those industries and occupations where such solvents are commonly used.

The responses of the nervous system to organic solvent toxicity can be classified as structural toxicity (general responses of the neuron and the supporting cells), biochemical toxicity (hypoxic or histotoxic changes, the effects of the metabolic products of the absorbed chemical), and the functional toxicity (sensory, motor and integrative functions). In most studies dealing with neurotoxicity only one or two of these aspects have been considered, which has, for obvious reasons, created much controversy owing to apparently conflicting results (e.g. Savolainen, 1982; Grasso et al, 1984; Spencer and Schaumburg, 1985). From the clinical point of view it is important to realize that individual susceptibility to solvents determines the clinical picture and the toxic responses following occupational exposure to solvents.

TABLE 1.

Solvent	Industry or occupation
Trichloroethylene and perchloroethylene	Decreasing, textile washing, and laundering
Styrene	Production of styrene polymers and reinforced plastics
Carbon disulfide	Viscose production
Toluene	Printing industry
n-Hexane	Shoe-making industry
Methyl-n-butyl ketone	Production of plastic coated fabrics
Mixture of solvents	Production of paints, varnishes and glues, the printing industry painting, varnishing, gluing cleaning and washing

SOME EPIDEMIOLOGIC ASPECTS

Epidemiologic evidence suggests that a risk for developing an disabling nervous system disease is increased among workers with occupational solvent exposure (Axelson et al, 1976; Axelson, 1983; Mikkelson, 1980; Olsen and Sabroe, 1980). However, when the literature on the effects of solvents on the nervous system is reviewed one is confused by the highly controversial results of different studies that have employed a wide spectrum of neurobehavioral methods (e.g. Hernberg, 1980; Friedlander and Hearne, 1980; Juntunen, 1983; Grasso et al, 1984). Questionnaires and a number of psychological and psychometric performance test batteries (Baker et al, 1983; Gamberale and Kjellberg, 1983; Lindström, 1982; Cherry et al, 1983; Cherry and Waldron, 1984;

Gamberale, 1985), sophisticated neurophysiological and other techniques (LeQuesne, 1982; Seppäläinen, 1985; Avanzini et al, 1983; Bleecker, 1983; Hagstadius and Risberg, 1983) have been employed in these studies.

The assessment of whether or not the differences in the functions of the nervous system between the exposed group and the referents with some of these methods are clinically meaningful is a very difficult task indeed. This difficulty is caused partly by semantic differences between investigators, partly by different selection of the groups of workers studied, and partly by the adaptive phenomena occurring in the exposed workers (e.g. Boyden, 1972). In particular, psychological studies are problematic: exposed and nonexposed individuals may not be equally well motivated for psychological testing; some diseased individuals tend to blame external factors for their poor health (partly due to compensational aspects involved); the selection of hypersusceptible individuals (Omenn, 1982) from among solvent-exposed groups; and many potential confounding and effect modifying factors such as alcohol, drugs and aging, just to mention a few of the problems involved. As many epidemiologists emphasize, the most common bias which can operate in epidemiologic studies is related to selection: if the criteria for a diagnosis include some requirement for exposure, any subsequent evaluation would result in an association between exposure and disease. This is a factor which should be seriously considered in modern highly developed societies, where occupational neurotoxic syndromes are mild and awareness about the risks of solvents both among phycicians and among workers has increased remarkably during the past few years (Hernberg, 1980; Axelson, 1983).

In this context, the important question of specificity and sensitivity of the examination techniques used in different clinical studies should be considered (e.g. Schoenberg, 1982; Juntunen, 1983). In this era of high technology, more sophisticated and sensitive techniques to detect early nervous system dysfunction are and will be available. Since the nonspecificity of the test increases along with increasing sensitivity, the relevance and validity of the new techniques must be very carefully tested with clinically well defined disease cases and healthy controls. Quite frequently, invalid clinical inferences have been drawn on the basis of small epidemiologic studies employing some new technique.

What, then, is the value of neurological examination in epidemiology of solvent poisoning? Some may believe that it is too laborious to be employed in large-scale epidemiologic studies and that it detects only relatively severe cases. Moreover, there may be problems in standardization of clinical tests. However, neurological examination is by no means more laborious than many of the commonly used examination techniques. It can also detect slight abnormalities, depending on the experience of the examiner. Standardization is also possible with regard to the clinical tests and recent development of quantitative neurology has greatly contributed to this. Comprehensive clinical examination and assessment of all the test results are necessary in individual diagnostics of solvent poisoning. This method yields the maximum specificity in its proper relationship to sensitivity in detecting subtle neurological disturbances.

CLINICAL PICTURE OF SOLVENT POISONING

From the clinical point of view, solvent poisonings can be classified into acute and chronic types. Acute intoxications are characterized by their dose-dependent narcotic effect on the central nervous system, which is also the basis for the medical uses of some of them in anesthesia. This effect is probably due to the fluidizing effect of the solvent on the nerve cell membranes. Acute effects are very common among workers exposed to solvents during their working day but severe acute intoxications are usually accidental. Disturbances in psychometric performances (e.g. Gamberale, 1985) and equilibrium (Savolainen and Linnavuo, 1979) have been shown to occur after a few hours of exposure to solvents, and these are in agreement with the clinical experience.

Chronic solvent poisoning refers to neurotoxic syndromes which have been described in association with long-term exposure to solvents. It is a highly controversial issue and has caused much debate in the literature (e.g. Grasso et al, 1984; Spencer and Schaumburg, 1985). The inherent problems in clinical, psychological and epidemiological study designs concerning exposure-effect relationship are familiar to everybody engaged in the research in this field. Many investigators still doubt whether a syndrome of chronic occupational solvent poisoning exists. On the basis of personal experience from about 2000 cases of suspected occupational solvent poisoning in Finland (all examined neurologically, neurophysiologically, psychologically and by other tests, see later) this author feels justified to state that chronic solvent poisoning does exist, although it may be considerably more infrequent than generally believed in some countries. In most cases the exposure data are inadequate and individual susceptibility to toxic agents makes it impossible to assess the exposure-effect relationships. The requirement of neuropathologic evidence in terms of structural alteration to prove neurotoxicity seems too strict in these cases and is, in fact, impossible to fulfil. There are only few types of specific solvent poisoning with specific neuropathologic features and associated clinical pictures.

Relatively few studies involving clinical examinations performed by neurologists have been carried out during the past decade. Toxic polyneuropathy due to methyl-n-butyl ketone has been reported in industrial workers (Allen et al, 1975; Spencer et al, 1975), due to carbon disulfide intoxication (Vigliani, 1954), to sniffing a glue thinner (Altenkirch et al, 1977), to a lacquer thinner (Means et al, 1976), to n-hexane (Cianchetti et al, 1976), to allyl chloride (He et al, 1980), to styrene (Lilis et al, 1978) and to trichloroethylene (Feldman, 1979). Although a number of solvents have been suspected of having this property, no convincing evidence supports this contention (Thomas, 1982). No appreciable clinical polyneuropathy has been found among car painters (Husman and Karli, 1980) or among patients with a diagnosed solvent poisoning (Juntunen et al, 1982a). A diagnosis of polyneuropathy is essentially a clinical one, and the diagnostic sensitivity depends on what symptoms and signs are considered sufficient to warrant the diagnosis of polyneuropathy. In borderline cases, electroneuromyography can confirm the diagnosis (Juntunen and Haltia, 1982).

Most solvents, however, exert their main actions on the central nervous system. One has to think of general neuropsychiatric effects (see e.g. Lipowski, 1980) when one is dealing with long-term solvent exposure. Early detection of the effects of chronic solvent exposure on the central nervous system is extremely difficult. Few clinical approaches to the problem have emphasized the frequent occurrence of a slight psycho-organic syndrome, motor and sensory disturbances and cerebellar dysfunction among workers exposed to organic solvents (Lilis et al, 1978; Arlien-Søborg et al, 1979, 1981; Husman and Karli, 1980; Struwe et al,

1980; Juntunen et al, 1980b, 1982a; Flodin et al, 1984). In attempts to associate solvent exposure to various neurological disorders, dementia and other psycho-organic syndromes, polyneuropathies and multiple sclerosis have been studied (Arlien-Søborg et al, 1979; Axelson et al, 1976; Mikkelsen, 1980; Amaducci et al, 1982). Brain atrophy is frequently associated with solvent exposure, but due to difficulties in study design, no convincing exposure-effect relationships have been established so far (Juntunen et al, 1980a). In some studies, exposed subjects even seem to have less atrophic changes than referents (Elofsson et al, 1980; Juntunen, 1985). Sophisticated analysis of the cerebrospinal fluid proteins and cells of patients exposed to organic solvents showed slight nonspecific alterations suggesting immunoactivation of the central nervous system (Juntunen et al, 1982b; Wikkelsø et al, 1984). Vestibular disturbances (Ödkvist et al, 1980; Arlien-Søborg et al, 1981; Binaschi and Cantu, 1982), disturbances in cerebral blood flow (Arlien-Søborg et al, 1982; Risberg and Hagstadius, 1983) and autonomic disturbances (Matikainen and Juntunen, 1982, 1985) have been described in association with exposure to solvents. Of particular interest is the clinical finding that cerebellar dysfunction and disturbances in gait and station are frequently encountered among patients: these functions can be examined only by clinical methods (Juntunen et al, 1982a). Long-term occupational exposure to toluene does not seem to affect appreciably the nervous system (Antti-Poika et al, 1985; Juntunen et al, 1985) while clear-cut cerebellar syndromes ensue among sniffers (e.g. Fornazzari et al, 1983).

The prognosis of organic solvent poisoning has been examined in some clinical studies. The results have been controversial: both improvement of subjective symptoms and signs (Bruhn et al, 1981), and deterioration of the signs after cessation of exposure (Antti-Poika et al, 1982; Juntunen et al, 1982a) have been reported. No clear-cut clinical picture of chronic organic solvent poisoning has emerged from these studies.

Of particular interest is the possibility of unexpected interactions of solvents and other exogenous factors, notably alcohol (Hills and Venable, 1982; Juntunen, 1982, 1984; Antti-Poika et al, 1985) and anaesthetics (Juntunen et al, 1984). In this field, the need for further scientific studies is evident.

DIAGNOSTIC CRITERIA FOR SOLVENT POISONING

Figure 1 demonstrates the flow-chart of the diagnostic procedure employed in individual diagnostics of solvent poisoning in Finland (Juntunen, 1982). When the suspicion of of organic solvent intoxication arises, the patient is admitted to the Institute of Occupational Health, Helsinki, for further examinations. The workers are usually well aware of the risks of occupational exposure and thus seek medical help rather readily and obvious selection of hypersusceptible individuals to examinations occurs. One inherent feature of the patients examined by an occupational neurologist is the mildness of the clinical picture. Whenever some objective findings suggesting nervous system dysfunction are observed, a careful differential diagnostics is perfomed. This is a crucial point in the diagnostic procedure and, unfortunately, seems to be inadequately performed in many scientific studies. If the etiology of the disease still remains uncertain, a comprehensive assessment of the whole case is performed, taking into account the data available on

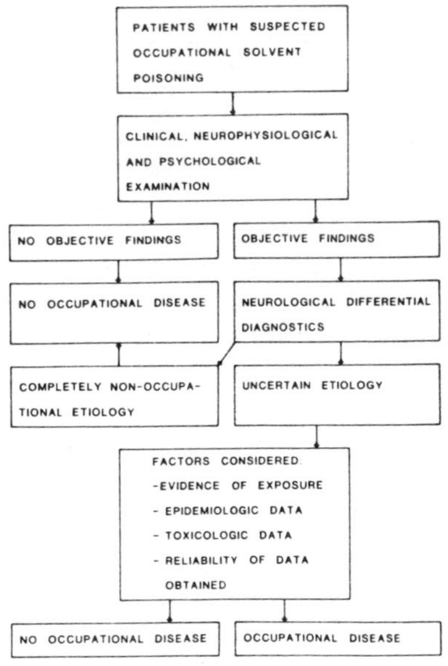

Fig. 1. Flow-chart of the diagnostic procedure for solvent poisoning (Juntunen, 1982).

exposure. The final diagnosis is always a probability diagnosis, and, in many cases a few months of follow-up is necessary to ascertain the diagnosis.

The following diagnostic criteria for solvent poisoning seem to be useful for practical purposes (Juntunen, 1978):

1) Verified relevant exposure to solvents known to be neurotoxic.
2) Clinical picture of organic damage to the nervous system:
   - typical subjective symptoms
   - objective findings in some of the following: clinical status, electroencephalography, electro-neuromyography, psychological tests
3) Other organic diseases reasonably well excluded (by neuroradiological methods, special laboratory tests or follow-up).
4) Primary psychiatric diseases reasonably well excluded.

These criteria have recently been accepted by the WHO expert meeting held in Copenhagen, and they will appear as WHO recommendations in near future.

SUMMARY AND CONCLUDINC REMARKS

If we accept that a syndrome of solvent poisoning exists, and I think there is substantial evidence to support this contention, we must also accept that there is a spectrum of clinical manifestations ranging from extremely mild cases to even fatal ones. The stage at which these mild syndromes are detected depends on individual susceptibility, on the examination techniques used, and on the prevailing socio-economical conditions in the community. Only few clinically relevant applications can be extracted from the bulk of data accumulated so far from different studies on solvent effects. This emphasizes the need for further studies and development of methods in quantitative neurology. Particularly, factors determining individual susceptibility provide a challenge for researchers in the field of occupational health.

REFERENCES

Allen N, Mendell JR, Billmaier DJ, Fontaine RE, O'Neill J (1975). Toxic polyneuropathy due to methyl n-butyl ketone. Arch Neurol 32: 209.

Altenkirch H, Mager J, Stoltenburg G, Helbrecht J (1977). Toxic polyneuropathy after sniffing a glue thinner. J Neurol 214: 137-156.

Amaducci L, Arfaioli C, Inzitari D, Marchi M (1982). Multiple sclerosis among shoe and leather workers: an epidemiological survey in Florence. Acta Neurol Scand 60: 94-103.

Antti-Poika M (1982). Overall prognosis of patients with diagnosed chronic organic solvent intoxication. Int Arch Occup Environ Health 51: 127-138.

Antti-Poika M, Juntunen J, Matikainen E, Suoranta H, Hänninen H, Seppäläinen AM, Liira J (1985). Occupational exposure to toluene: neurotoxic effects with special emphasis on drinking habits. Int Arch Occup Environ Health 56: 31-40.

Arlien-Søborg P, Bruhn P, Gyldensted C, Melgaard B (1979). Chronic painters' syndrome. Chronic toxic encephalopathy in house painters. Acta Neurol Scand 60: 149-156.

Arlien-Søborg P, Henriksen L, Gade A, Gyldensted C, Paulson O (1982). Cerebral blood flow in chronic toxic encephalopathy in house painters exposed to organic solvents. Acta Neurol Scand 66: 34-41.

Arlien-Søborg P, Zilstorff K, Grandjean B, Pedersen M (1981). Vestibular dysfunction in occupational chronic solvent intoxication. Clin Otolaryngol 6: 285-290.

Avanzini M, Casazza M, Girotti F (1983). Electro-oculography. In Gilioli R, Cassitto MG, Foa V (eds) "Neurobehavioral Methods in Occupational Health." Advances in the Biosciences, vol 46. Pergamon Press, pp 47-53.

Axelson O (1983). Epidemiology of solvent related neuropsychiatric disorders. In Cherry N, Waldron HA (eds) "The Neuropsychological Effects of Solvent Exposure." The Colt Foundation, pp 85-99.

Axelson O, Hane M, Hogstedt C (1976). A case-referent study on neuro-psychiatric disorders among workers exposed to solvents. Scand J Work Environ Health 2: 14-20.

Baker EL, Feldman RG, White RF, Harley JP, Dinse GE, Berkey CS (1983). Monitoring neurotoxins in industry: development of a neurobehavioral test battery. J Occup Med 25: 125-130.

Binaschi S, Cantu L (1983). Vestibular function in solvent exposure: clinical criteria. In Gilioli R, Cassitto MG, Foa V (eds) "Neurobehavioral Methods in Occupational Health." Advances in the Biosciences, vol 46. Pergamon Press, pp 205-210.

Bleecker ML (1983). Optacon - a new screening device for peripheral neuropathy. In Gilioli R, Cassitto MG, Foa V (eds) "Neurobehavioral Methods in Occupational Health." Advances in the Biosciences, vol 46. Pergamon Press, pp 41-46.

Boyden S (1972). The environment and human health. Med J Australia 1: 1229-1234.

Bruhn P, Arlien-Søborg P, Gyldensted C, Christensen EL (1981). Prognosis in chronic toxic encephalopathy. A two years follow-up study in 26 house-painters with occupational encephalopathy. Acta Neurol Scand 64: 259-272.

Cherry N, Venables H, Waldron HA (1983). The use of reaction time in solvent exposure. In Gilioli R, Cassitto MG, Foa V (eds) "Neurobehavioral Methods in Occupational Health." Advances in the Biosciences, vol 46. Pergamon Press, pp 191-195.

Cherry N, Waldron H (1984). "The Neuropsychological Effects of Solvent Exposure." The Colt Foundation.

Cianchetti C, Abritti G, Perticoni G, Siracusa A, Curradi F (1976). Toxic polyneuropathy of shoe-industry workers. J Neurol Neurosurg Psychiatr 39: 1151-1161.

Eloffson SA, Gamberale F, Hindmarsch T, Iregren A, Isaksson A, Johnsson I, Knave B, Lydahl E, Mindus P, Persson HE, Philipsson B, Steby M, Struwe G, Söderman E, Wennberg A, Widen L (1980). A cross-sectional epidemiologic investigation on occupationally exposed car and industrial spray painters with special reference to the nervous system. Scand J Work Environ Health 6: 239-273.

Feldman RG (1979). Trichloroethylene. In Vinken PJ, Bruyn GW (eds) " Intoxications of the Nervous System," part 1, vol 36, North-Holland Publishing Company, pp 457-468.

Fiserova-Bergerova V (1985). Toxicokinetics of organic solvents. Scand J Work Environ Health 11:suppl 1, 7-21.

Flodin V, Edling C, Axelson O (1984). Clinical studies of psycho-organic syndromes among workers with exposure to solvents. Am J Ind Med 5: 278-295.

Fornazzari L, Wilkinson DA, Kapur BM, Carlen PL (1983). Cerebellar, cortical and functional impairment in toluene abusers. Acta Neurol Scand 67: 319-329.

Friedlander BR, Hearne FT (1980). Epidemiologic considerations in studying neurotoxic disorders. In Spencer PS, Schaumburg HH (eds) "Experimental and Clinical Neurotoxicology." Baltimore, London: Williams and Wilkins, pp 650-662.

Gamberale F (1985). Use of behavioral performance tests in the assessment of solvent toxicity. Scand J Work Environ Health 11:suppl 1, 65-74.

Gamberale F, Kjellberg A (1983). Field studies of the acute effects of exposure to solvents. In Cherry N, Waldron HA (eds) "The Neuropsychological Effects of Solvent Exposure." The Colt Foundation, pp 117-129.

Grasso P, Sharrott M, Davies DM, Irvine D (1984). Neurophysiological and psychological disorders and occupational exposure to organic solvents. Food Chem Toxicol 22: 819-852.

Hagstadius S, Risberg J (1983). Regional cerebral blood flow in subjects occupationally exposed to organic solvents. In Gilioli R, Cassitto MG, Foa V (eds) "Neurobehavioral Methods in Occupational Health." Advances in the Biosciences, vol 46. Pergamon Press, pp 211-217.

He F, Shen D, Guo Y, Lu B (1980). Toxic polyneuropathy due to chronic allyl chloride intoxication. A clinical and experimental study. Chinese Med J 93: 177-182.

Hernberg S (1980). Neurotoxic effects of long-term exposure to organic hydrocarbon solvents. Epidemiologic aspects. In Holmstedt B, Lauwerys R, Mercier M, Roberfroid M (eds) "Mechanisms of Toxicity and Hazard Evaluation." Elsevier/North-Holland Biomedical Press.

Hills BW, Venable HL (1982). The interaction of ethyl alcohol and industrial chemicals. Am J Ind Med 3: 321-333.

Husman K, Karli P (1980). Clinical neurological findings among car painters exposed to a mixture of organic solvents. Scand J Work Environ Health 6: 33-39.

Juntunen J (1978). Neurotoxic syndromes in man. 3rd International Course in Industrial Toxicology. Helsinki: Institute of Occupational Health, pp 164-168.

Juntunen J (1982). Alcoholism in occupational neurology: diagnostic difficulties with special reference to the neurological syndromes caused by exposure to organic solvents. In Juntunen J (ed) "Occupational Neurology." Acta Neurol Scand 66:suppl 92, 89-108.

Juntunen J (1983). Neurological examination and assessment of the syndromes caused by exposure to neurotoxic agents. In Gilioli R, Cassitto MG, Foa V (eds) "Neurobehavioral Methods in Occupational Health." Advances in the Biosciences, vol 46. Pergamon Press, pp 3-10.

Juntunen J (1984). Alcohol, work and the nervous system. Scand J Work Environ Health 10: 461-465.

Juntunen J, Antti-Poika M, Tola S, Partanen T (1982a). Clinical prognosis of patients with diagnosed chronic organic solvent intoxication. Acta Neurol Scand 65: 488-503.

Juntunen J, Haltia M (1982). Polyneuropathies in occupational neurology: pathogenetic and clinical aspects. In Juntunen J (ed) "Occupational Neurology." Acta Neurol Scand 66:suppl 92, 59-73.

Juntunen J, Hernberg S, Eistola P, Hupli V (1980a). Exposure to industrial solvents and brain atropy. European Neurology 19: 366-375.

Juntunen J, Hupli V, Hernberg S, Luisto M (1980b). Neurological picture of organic solvent poisoning in industry. Int Arch Occup Environ Health 46: 219-231.

Juntunen J, Kaste M, Härkönen H (1984). Cerebral convulsion after enfluran anaesthesia and occupational exposure to tetrahydrofuran. J Neurol Neurosurg Psych 47: 1258.

Juntunen J, Matikainen E, Antti-Poika M, Suoranta H, Valle M (1985). Nervous system effects of long-term occupational exposure to toluene. Acta Neurol Scand (in press).

Juntunen J, Taskinen E, Luisto M, Iivanainen M, Nurminen M (1982b). Cerebrospinal fluid cells and proteins in patients occupationally exposed to organic solvents. J Neurol Sci 54: 413-425.

Järvisalo J, Tossavainen A (1982). Exposure to neurotoxic agents in Finnish working environments. Trends and assessment of exposure. In Juntunen J (ed) "Occupational Neurology." Acta Neurol Scand 66:suppl 92, 37-45.

LeQuesne PM (1982). Electrophysiological investigation of toxic neuropathies. In Juntunen J (ed) "Occupational Neurology." Acta Neurol Scand 66:suppl 92, 75-87.

Lilis R, Lorimer WV, Diamond S, Selikoff IJ (1978). Neurotoxicity of styrene in production and polymerization workers. Environ Res 15: 133-138.

Lindström K (1982). Behavioral effects of long-term exposure to organic solvents. In Juntunen J (ed) "Occupational Neurology." Acta Neurol Scand 66:suppl 92, 131-141.

Lipowski Z (1980). Organic mental disorders; Introduction and review of syndromes. In Kaplan HI, Freedman AM, Sadock BJ (eds) "Comprehensive Textbook of Psychiatry/III." Baltimore, London: Williams and Wilkins.

Matikainen E, Juntunen J (1982). Chronic solvent exposure and autonomic nervous system involvement. Acta Neurol Scand 65:suppl 90, 230-231.

Matikainen E, Juntunen J (1985). Autonomic nervous system dysfunction in workers exposed to organic solvents. J Neurol Neurosurg Psych 48: 1021-1024.
Means ED, Prockop LD, Hooper GS (1976). Pathology of lacquer thinner induced neuropathy. Ann Clin Lab Sci 3: 240-250.
Mikkelsen S (1980). A cohort study of disability pension and death among painters with special regard to disabling presenile dementia as an occupational disease. Scand J Soc Med, suppl 16: 34-43.
Olsen J, Sabroe S (1980). A case-referent study of neuro-psychiatric disorders among workers exposed to solvents in the Danish wood and furniture industry. Scand J Soc Med, suppl 16: 44-49.
Omenn GS (1982). Predictive indentification of hypersusceptible individuals. J Occ Med 24: 369-374.
Risberg J, Hagstadius S (1983). Effects on the regional cerebral blood flow of long-term exposure to organic solvents. Acta Psychiat Scand 67: 92-99.
Savolainen H (1982). Toxicological mechanisms in acute and chronic nervous system degeneration. In Juntunen J (ed) "Occupational Neurology." Acta Neurol Scand 66:suppl 92, 23-35.
Savolainen K, Linnavuo M (1979). Effects of m-xylene on human equilibrium measured with quantitative method. Acta Pharmacol Toxicol 44: 315-318.
Schoenberg BS (1982). Descriptive neuroepidemiology: applications to occupational neurology. In Juntunen J (ed) "Occupational Neurology." Acta Neurol Scand 66:suppl 92, 1-9.
Seppäläinen AM (1982). Neurophysiological findings among workers exposed to organic solvents. In Juntunen J (ed) "Occupational Neurology." Acta Neurol Scand 66:suppl 92, 109-116.
Spencer PS, Schaumburg HH (1985). Organic solvent neurotoxicity facts and research needs. Scand J Work Environ Health 11:suppl 1, 53-60.
Spencer PS, Schaumburg HH, Raleigh RL, Terhaar CJ (1975). Nervous system degeneration produced by the industrial solvent methyl n-butyl ketone. Arch Neurol 32: 219-222.
Struwe G, Mindus P, Jonsson B (1980). Psychiatric ratings in occupational health research: a study of mental symptoms in lacquerers. Am J Ind Med 1: 23-30.
Thomas PK (1980). The peripheral nervous system as a target for toxic substances. In Spencer PS, Schaumburg HH (eds) "Experimental and Clinical Neurotoxicology." Baltimore, London: Williams and Wilkins, pp 35-47.

Vigliani EC (1954). Carbon disulphide poisoning in viscose rayon factories. Br J Ind Med 11: 235-247.
Wikkelsø C, Ekberg K, Lilienberg L, Wettenholm B, Karlsson B, Blomstrand C, Johansson B (1984). Cerebrospinal fluid proteins and cells in men subjected to long-term exposure to organic solvents. Acta Neurol Scand 70:suppl 100, 113-119.
Ödkvist L, Larsby B, Fredrickson J, Liedgren S, Tham R (1980). Vestibular and oculomotor disturbances caused by industrial solvent. J Otolaryng 9: 53-59.

PREVENTION AND PROTECTION

Safety and Health Aspects of Organic Solvents, pages 283–296

DANISH WORK ENVIRONMENT REGULATION OF ORGANIC SOLVENTS: AN ATTEMPT TO EVALUATE THE EFFECTS.

Ole Svane
Danish Labour Inspection Service
Landskronagade 33-35
DK 2100 Copenhagen Ø
Denmark

A. Historical Trend 1945-1980

At the end of the Second World War a strong wave of public health consciousness went over the Scandinavian countries. In 1945 the Danish National Inspection Services published a short list of exposure limits as the Americans did in 1946. In this list 30 limit values were assigned to solvents on the basis of toxicological studies and experiences from several countries, especially Germany.

In Table 1 a few examples of exposure values are shown. The general impression is that the recommendations were quite cautious concerning aromatic carbohydrates already in 1945. The exposure values of other solvents have been lowered considerably during the following forty years.

During the fifties and sixties no new exposure values were published in Denmark. Foreign guidelines were used as recommendations. Specific work tasks were regulated or guidelines were published by the authorities. Examples are metal degreasing and polyester moulding.

Following the debate in 1972 on the first Danish report on chronic toxic encephalopathy in house painters the union, the employers and the manufacturers of paints gathered and began the work of elaborating a labelling system. The paints code agreed on by the three parties and the Labour Inspection Services was the forerunner of the present paints code included in the 1980/82 regulation of paints.

An overall view gives the impression of an awareness of the risks of solvent exposure rising during a post-war 30 year period. The public authority's reaction was a mixture of recommendations, guidelines, a few specific orders and an agreement on the paints code.

Table 1

Historical Background of the Present Regulation of Organic Solven

HISTORICAL BACKGROUND

1930ies German Exposure Values

1940ies Scandinavian and American Exposure Values

	1945	1985
	mg/m^3	
Chloroform	100	10
Toluene	200	280
Acetone	1000	600

1950ies Guidelines and Regulations:
Metal degreasing

1960ies Dry cleaning
Polyester moulding
Glueing and priming in building industry

1970ies Chr. Toxic Encephalopathy debate
Tripartite Paints Code

B. The 1980/82 Regulation

In 1980 the Minister of Labour was so impressed by the amount of research and still more dramatic case stories concerning demented house painters that he asked the Directorate of the Labour Inspection to impose a ban on the solvents based house paints. Even though he promised the partners who were both present in the radio studio where his decision was taken, that the regulation would be in force after a three-months period, it took 18 months of hard work to elaborate the rules which have now been enforced three years ago.
First of all the regulation is based on a ban. The ban is taken literally on two heavy exposure situations: painting large surfaces, i.e. indoor walls and indoor ceilings. The ban is modified in other situations, for example outdoor painting and painting small surfaces.

By elaborating a code system which was a continuation of the agreement code system from the seventies we were able to impose a mandatory substitution. It means in plain words that you have to find the work task that you want to perform in a small booklet containing the order and a guidance note. Thereby you can look up the code which is the maximum allowable code.

The code consists of two figures (Table 2). The figure before the hyphen denotes the health risk of the vapours emanating from the product and the figure after gives an account of the risks of the non-vapourized part of the products.

Let us take an example: A paints product consists of three solvents.

By a simple additive model the risks of three solvents are combined into one single figure - the figure before the hyphen. You must know the exposure limit value, the relative evaporation rate (in relation to n-butylacetate) and the weight per cent of each of the solvents. In that way the code figure before the hyphen divides the products into seven classes.

A manufacturer, importer or dealer who informs about certain proportions of mixtures of components, adding of diluents, etc. must also indicate the code number for the ready-for-use mixture.

Table 2

An example of a label showing the code number.

Attention is drawn to the fact:

- that the figure before the hyphen will often be higher if a product is diluted with organic solvents,
- that a cleaning agent to which water is added the ready-for-use mixture usually will have a lower figure before the hyphen than the cleaning agent in the unmixed state,

- that in certain multi-component systems after mixing of the components there will be less monomers left partly because of dilution of the components and partly because of reaction between the components. This may result in a lower code number for the final mixture than for the individual components,

- that in certain multi-component systems the mixing may result in release of volatile components which may lead to higher code number for the final preparation than for the individual components.

The figure after the hyphen in the code number expresses the health risk.

1) if skin and eyes come into direct contact with the product,
2) by inhalation of drops or dust from a spray mist of the product or dust released from the product,
3) by ingestion of the product (for example in connection with eating or smoking).

The figure after the hyphen divides the products into six groups.

The figure after the hyphen is determined taking into account all components which make up the product.

In Table 3 a short description of the risks in the groups determined by the figure after the hyphen is given.

A set of decision rules are laid down in the order of determining code numbers by which it is possible to include the relative weight of the risks of all components of the product.

The figure before the hyphen divides the products into seven classes and the figure after the hyphen divides the products

Table 3

Product groups and their risks. The groups determined by the figure after the hyphen.

Description of risks in the groups	The figure after the hyphen
Water	-0
Products containing components with no known hazards in connection with non-dirty painting work, but with hazards in connection with inhalation of spraying mist, dust etc.	-1
Products containing components with hazards in connection with ingestion and inhalation of spraying mist, dust etc. but without known hazards to skin and eyes in connection with non-dirty painting work	-2
Products containing components which may be hazardous in connection with contact with skin and eyes and in connection with inhalation of spraying mist, dust etc. The hazard may also be allergy	-3
Products containing components which may involve a risk of corrosive attacks on the body	-4
Products containing components with strong allergic effects in connection with skin contact or which are especially hazardous in contact with skin and eyes	-5
Products containing components which are toxic in connection with contact with skin and eyes and in connection with inhalation of spraying mist, dust etc. and when ingested in small quantities	-6

into six groups. This means that the characterization is very detailed, but on the other side it is a non-specific toxicological information. Comparing with warning systems of the European Communities, the EEC Risk Sentences give specific information on what sort of risk the consumer may run - if the system is used according to its ideal intentions. The Danish labels carry both types of information.

In the Danish code system practical toxicologists have made choices on behalf of the painter. This means that information is simple and that the message goes over: choose the lowest code possible to guard your own health. The immediate effect of the code system is therefore a basis for a more healthy choice when buying the product in the paints shop.

The second effect of labelling with code numbers is restricting the choice for certain processes. The regulation contains for example detailed instructions on the choice for each process detail of painting window frames on the inside surface. The code mentioned is maximally permitted unless a certain reason can be documented.

The third element of the regulation is the detailed description of work practices and safety measures for each of the painting procedures. It contains demands on ventilation, safety measures, working clothes, preparation of surfaces, work place design, and personal hygiene.

C. The Effect of the Regulation: Dream or Reality?

The dream of the National Labour Inspection is of course a painter sitting on his container of paints, reading our instructions. A large program of marketing the code system and the recommendations has been the result of many peoples' work in the National Labour Inspection Services and the Work Environment Fund.

But the basis of the real hightening of consciousness in the public was the unanimous agreement between the partners of the labour market and the producers with the National Labour Inspectorate on the seriousnes of the problem. One of the explanations was that many house painters, who were now employers, had been working hard in their own small enterprises and work-shops and had experienced the acute and chronic effects of solvents based paints.

I should also mention that the present activities of the Labour Inspection Services on solvents are still very vigorous. This year we have conducted four campaigns in four different branches (wood and furniture, anti-rust car coating, metal degreasing and electronics industry), which means that all shops, enterprises and companies have been inspected during a three months period with a special check list and a specially designed follow-up program. It means that the National Labour Inspection makes an example of employers not obeying the rules by taking them into court. The public consciousness and debate is very high at present, which is of course mainly a result of the research performed in Denmark and in other countries.

It is often discussed if you are able to trace the effects of the preventive work done by public authorities. This question has also been raised on the paints regulations.

The final effect, namely the elimination or diminishing of acute as well as chronic toxic effects of solvents, should be possible to detect after a couple of years.

An intermediate effect is a rise of the hygiene in work places. Some of the parameters to make an assessment of that kind are present.

Looking at the figures on net consumption (supply) of some widely used solvents, Table 4, you may get the impression that some change has taken place in the period 1970-1984. A total account of all solvents is needed to evaluate the tendency in the table, and a distribution on different sorts of consumption would have been informative. Especially the activity in the building and construction industry might influence the consumption of paints.

By permission of the Danish Paints Industry an account of the total Danish consumption of paints 1975-83 is shown in Table 5. The columns describe the total consumption with the percentage of latex paints that contain only a few per cent of volatile organic solvents. The circles give an impression of the percentage of house paints that are water based. The tendency is clearly in favour of non-solvents based paints during the last ten years.

The Danish Labour Inspectorate has tried to evaluate the effect of the recent regulation by setting up a small survey

Table 4.

Total net Consumption (supply) of some Solvents in Denmark 1970-84.

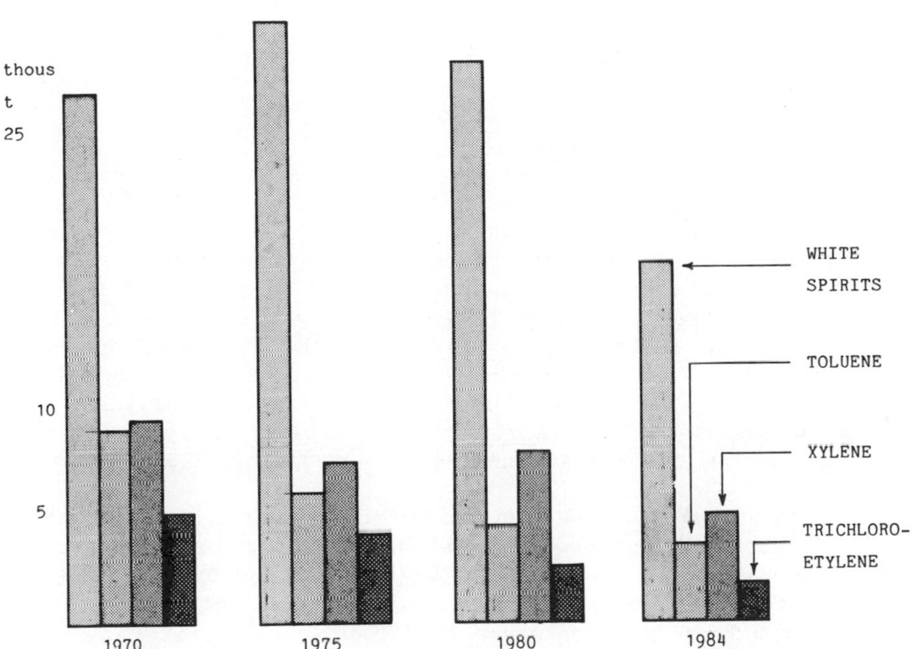

(Source: Danish National Bureau of Statistics).

with some elements of research. The method was an inspection and interview of a stratified sample of house painters' work shops. In the shops or workplaces a simple sample of persons - either employers or employees - were interviewed on their knowledge of the regulation, and of their work practices. The choice of products and the safety measures were assessed by the inspectors and compared with the demands of the regulations. You may of course have some methodological considerations on the possible information bias by using inspectors as interviewers and observers. The inspectors are people who have been skilled workers in the building industry for years.

Only a few results can be shown here. In a large region 101 house painters' work were inspected. In Table 6 it appears that 13 of 101 shops did not choose paints products according to the rules. The answers on a question on their knowledge of the regulation of paints do not look flattering when distributed according to their factual product choice. The total impression is, however, that the choice of products was more favourable than the maximally permitted code, and work practices lived up to the regulations and guidelines. Perhaps the most important factor is the supplier of paints, i.e. what is on the shelves in the shop where the painters buy.

If you turn to the 'real' effect side, i.e. information on diseases or accidents, there is no trend in accidents in the the period 1974-84. This is well-explained by the deficiencies of the reporting rate, as you may imagine only about thirty per cent of all accidents are notified.

Table 7 shows an account of the cases of chronic toxic encephalopathy accepted by the National Insurance Board. The tendency is a rise during the period 1977-1983. Analysis of cases cannot be performed by conventional methods as data are not accessible in a registry. A preliminary analysis of the cases acepted 1977-81 shows that painters, printers and metalworkers make up the largest part of cases.

The National Labour Inspection Services has introduced a new registry of cases notified to the two public authorities, i.e. the National Labour Inspection and the National Insurance Board. The cases are not verified but suspected by the notifying physicians. In 1983 1010 cases were notified. Persons working in painting offices, serigraphic work shops, house and construction workers, and car painters are the most commonly mentioned patients.

Table 5.

Danish Paints Consumption 1975-83.

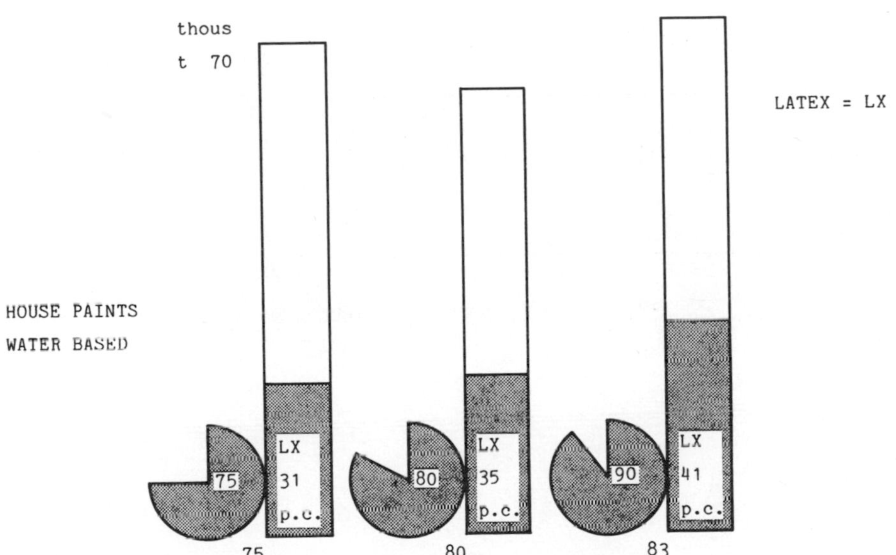

(Source: Danish Paints Industry Association).

Table 6

The Choice of Products. Interview and Control of a Stratified Sample of Hou
Painters 1984.

Know Booklet?	Product Choice OK or Better	Worse	Total
Yes	61	10	71
No	27	3	30
Total	88	13	101

(Source: National Labour Inspection, Statistical Services)

Table 7

Cases of Chronic Toxic Encephalopathy Accepted or Notified in Denmark.

I. NATIONAL INSURANCE BOARD

Year of acceptance	Number of cases
1977	15
1978	28
1979	45
1980	174
1981	205
1982	358
1983	593

II. NOTIFIED NUMBER OF CASES

1983	(Chronic) Toxic Encephalopathy	1010

Note: Danish Population 5 mill.
Danish work force 2.8 mill.

D. Future outlook

The experiences of the paints code regulation has been so positively received by all partners on the working market that the Ministry of Labour has asked the Labour Inspectorate to enlarge the area regulated. The Inspectorate is now preparing a code system for products such as glues, degreasing agents and printing agents. It is also under consideration to start a notification of all products containing solvents.

Furthermore the paints regulation will continuously be revised, especially concerning the restricted choice. The development of new products has facilitated the demand of still lower code numbers for certain processes.

Looking back, my personal judgement is positive: A long trend of rising attention has been accompanied by guidelines, recommendations and new orders. The orders did not make it: it was the acceptance of the problem that gave the good hygienic results. But the regulation is necessary to keep the train on the right track.

The full text of the Order on Professional Painting Work and the Order of Determining Code Numbers is available in English.

Safety and Health Aspects of Organic Solvents, pages 297–302

THE SYSTEMATIC CHOICE OF MATERIAL IN PERSONAL PROTECTION EQUIPMENT FOR ORGANIC SOLVENTS

Charles M. Hansen

Scandinavian Paint and Printing Ink Research Institute, Agern Allé 3, DK-2970 Hørsholm, Current address: Jens Bornøsvej 16, DK-2970 Hørsholm, Denmark.

INTRODUCTION

There is considerable current research to eliminate or reduce levels of organic solvents in many products where they previously were used in large quantities. Water is the preferred substitute but for technical reasons its use has not been universally possible. Reformulation to less hazardous solvents has also been done in many instances, but this is not always possible either. In situations where organic solvents are still encountered and health dangers do exist, one must still resort to ventilation and personal protection devices. This report will concentrate on recent applications of the solubility parameter as a tool in choosing appropriate materials for personal protection and demonstrate its potential for broader applications in the field of occupational health and safety.

THE SOLUBILITY PARAMETER AND MEMBRANE PENETRATION

The permeation rate of various penetrants through polyvinyl chloride glove membranes has been correlated by Henriksen using solubility parameters (Henriksen, 1982). There is an exponential increase in the permeation rate with a variation of a factor of 1000 in the rate. In other words a "good" solvent, can penetrate 1000 times faster than a "bad" one. There was some scatter in the correlation probably due to the differences in size and shape of the penetrants. Henriksen's report, which also includes similar data on other membrane types, has been translated into English. There is considerable interest in this area in Scandinavia, perhaps because of the medical reports linking solvent exposure to brain damage which are regularly seen on television.

These Scandinavian efforts to correlate solubility parameters with penetration of protective clothing by organic materials are based on previous Scandinavian research. A doctoral thesis from 1967 includes the background for these developments (Hansen 1967a).

The penetration of membranes depends on both the solubility, S, and the diffusion coefficient, D. The usual equation for the permeation rate, P, is

$$P = DS$$

An increase in solubility leads to a proportional increase in the permeation rate, for D being the same. Solubility correlates with the solubility parameter. It has been shown (Hansen 1967a, b) that the diffusion coefficient in polymer films varies exponentially with organic penetrant concentration. This is general behavior for penetrant-polymer systems. The penetrant plasticizes more as its concentration increases and the mobility (diffusion coefficient) in the system increases. Thus, the solubility and hence the solubility parameter also play a role in the diffusion rate because of these effects. Diffusion rates were found to be higher for smaller and more linear molecules. Space requirements rather than physical affinities determined the diffusion rates once the solubility allowed the presence of the penetrant in the films.

The most widely used solubility parameter concept today is the extension of Hildebrand's concepts attributable to Hansen (Hansen 1967a). A CRC handbook by Barton is recommended as the most recent comprehensive treatment of the field in general (Barton, 1983).

Hansen divided the cohesion energy into three parts to account for the three independent energy types commonly called dispersion (London), permanent dipole-permanent dipole, and hydrogen bonding. These are the major energy types which cause interactions between organic materials.

When these energies are suitably well matched for two materials, solubility occurs. Solubility regions can be defined. In cross-linked polymeric materials this matching of energies causes swelling by penetrant. The first correlation of swelling of cross-linked polymers with these solubility parameters was by Beerbower and Dickey (Beerbower, Dickey 1969). Beerbower later collaborated with Hansen in a general review on solubility parameters (Hansen, Beerbower, 1971). In general the better the solvent quality the greater the polymer swelling up to a maximum. Solvent quality can, of course, be estimated by solubility parameter techniques.

SOLUBILITY PARAMETERS FOR POLYMERS IN PROTECTIVE CLOTHING

A present problem of some dimension is that only a relatively few polymeric types used in protective clothing are characterized by these techniques. These are to be found in the publications by Hansen and Henriksen cited earlier as well as in a report from the Scandinavian Paint and Printing Ink Research Institute (Saarnak, Hansen, 1982).

The use of group contribution methods to calculate solubility parameter relations for polymers is very questionable. There are far too many parameters influencing behavior to allow relatively simple calculations having a respectable degree of reliability.

A warning is in order at this point. What is generally called polyethylene in this trade is in fact a polyethylene product modified, for example, with other monomers to improve laminating properties or other properties. Data on polyethylene as such, or any other polymer as such, can be misleading. One should really examine the membrane materials in question. It can be said from experience, however, that the number of solvents having affinity for a given polymer type is not reduced by altering its structure i.e. by including co-monomers. In other words the solubility parameter regions are generally made larger by co-polymerizing.

USE OF POLYMER SOLUBILITY PARAMETER DATA FOR PROTECTIVE CLOTHING

Several recent discussions with the researchers in this area in Denmark have led to the following current view of how to use solubility parameter data for polymers in selecting protective clothing.

If the affinity of the penetrant is such that it is within the region characterizing the polymer, then this polymer is not to be selected for protective purposes unless very critical evaluation of thicknesses etc. clearly shows that it can be used. Such a high affinity must generally indicate an unsuitable combination. Thus rapid screening of unsuited combinations is possible.

On the other hand, the fact that the affinities are low enough to place the penetrant outside the solubility (specified percent swell) range does not mean the combination can be used without critical review. Water, for example penetrates polyethylene to some extent. Their solubility parameters are widely different; these have very low affinities for each other. No membrane can be considered perfect for use forever, and the breakthrough times and permeation rates which can be tolerated will depend on the given situation. In general, however, the larger the differences between the penetrant and polymer solubility parameters, the more secure the system will be.

An important extension to this lack of affinity leading to improved barrier properties is as follows: If data are available on given (model) penetrant/polymer combinations indicating that penetration is retarded to a safe level for a given purpose, then a penetrant having simular molecular size and shape (i.e. approximately the same diffusion coefficients (Hansen, 1967a) which has lesser affinity for the polymer should also be satisfactory. The relative toxicological effects of the penetrant in question must also be considered, of course.

This proposal is important in many situations where no data are available on penetration characteristics for a given penetrant and a quick decision must be made as to a suitable protective membrane.

SOLUBILITY PARAMETER CALCULATIONS

When the solubility parameters for two materials are known, one can calculate an energy difference which reflects their affinities for each other. The equation which has been used for this purpose for many years is:

$$R_A^2 = 4(\delta_{D_1} - \delta_{D_2})^2 + (\delta_{P_1} - \delta_{P_2})^2 + (\delta_{H_1} - \delta_{H_2})^2$$

where the subscripts 1 and 2 denote the materials involved, and R_A is the energy difference between them in solubility parameter space. The 4 enters as an empirical factor to allow data to be plotted as spherical regions of solubility. This allows easier computer treatment of data as well.

R_A is then compared with the reported data for the solutes or polymers which are given in tables or experimentally determined as a characteristic radius, R. If R_A is less than R for a polymer, solubility is predicted. The value R_A/R can be used as a relative energy difference parameter (RED number). For RED numbers of 0 there is a perfect match of energies. The solubility parameters are identical. RED number of 1.0 means a boundary solvent and higher RED numbers mean greater energy differences and lower affinities. This has been an easier concept to master for those who are not technically trained.

THE SOLUBILITY PARAMETER AND BIOLOGICAL SYSTEMS

The affinities of organic solvents in biological systems have been recently described using solubility parameter techniques (Hansen, Andersen, 1985). Lard, water, blood serum, sucrose, urea, keratin, and lignin were studied. The tendency for organic solvents to collect in fatty material, swell (penetrate) the skin, absorb into wood, or preferentially transfer to aqueous media, for example, can be estimated rapidly for any of the solvents for which solubility parameters are available. The solubility parameter data on these materials are included in Table 1. The skin (membrane) penetration as evaluated from the swelling of psoriasis scales (Keratin) was treated in detail in an earlier report (Hansen, 1982). The solubility parameter can thus be used to estimate affinities in a large variety of systems where differences in behavior are found for a suitably large number of organic solvents.

It is hoped these correlations and the possibilities they offer to correlate other phenomena, will give inspiration for additional research and applications. It would appear that a total systems analysis for the physical behavior of organic solvents is possible in many systems of interest in occupational health and safety.

Table 1 Solubility Parameter Data for Some Biologically Interesting Materials $(J/cm^3)^{\frac{1}{2}}$

		δ_D	δ_P	δ_H		DTOT
A	Fat(lard) 37°	15.90	1.16	5.41	12.03	1.000
B	Fat(lard) 23°	17.69	2.66	4.36	7.98	1.000
C	1% in water*	15.07	20.44	16.50	18.12	0.856
D	1% in water	14.96	18.33	15.15	16.22	0.801
E	Blood serum	23.20	22.73	30.60	20.48	0.955
F	Sucrose	21.67	26.26	29.62	20.44	0.956
G	Urea	20.90	18.70	26.40	19.42	0.983
H	Psoriasis scales	24.64	11.94	12.92	19.04	0.927
I	Lignin	20.61	13.88	15.25	11.83	0.964

* Amines not included - preferred values
DTOT is a measure of the degree of fit of the data.

REFERENCES

Barton AFM (1983). Handbook of Solubility Parameters and other Cohesion Parameters. CRC Press Inc Boca Raton Florida.

Beerbower A, Dickey JR (1969). Advanced Methods for Predicting Elastomer/Fluid Interactions. ASLE Transactions 12:1.

Hansen CM (1967a). Doctoral Dissertation - The Three Dimensional Solubility Parameter, Danish Technical Press, Copenhagen.

Hansen CM (1967b). The Measurement of Concentration Dependent Diffusion Coefficients. IND ENG CHEM Fundamentals 6:No.4, 609.

Hansen CM (1982). The Absorption of Liquids into the Skin. Report T 3-82, Scand. Paint Printing Ink Res. Inst.

Hansen CM, Andersen BH (1985). The Affinities of Organic Solvents in Biological Systems. Report T 1-85, Scand. Paint Printing Ink Res. Inst.

Hansen CM, Beerbower A (1971). Solubility Parameters. Encyclopedia of Chemical Technology Supplement Volume 2nd Ed., Wiley New York 889.

Henriksen HR (1982). Selection of Materials for Protective Gloves. Polymer Membranes to Protect Against Contact with Epoxy Products. Report No. 8/1982. Danish National Labour Inspection Service, Copenhagen (Translation in English available).

Saarnak A, Hansen CM (1982). Solubility Parameters - Characterization of Coatings, Binders, and Polymers. Report T 10-82. Scand. Paint Printing Ink Res. Inst.

Safety and Health Aspects of Organic Solvents, pages 303–315

PERSONAL PROTECTION WHEN HANDLING SOLVENTS

Juhani Jaakkola

Department of Industrial Hygiene and Toxicology, Institute of Occupational Health, SF-00290 Helsinki, Finland

INTRODUCTION

The use of respiratory protective devices and protective gloves is necessary if technical control measures and appropriate working practices alone cannot safeguard the worker against hazardous exposure. Typical operations and industries where respirators and protective gloves are frequently needed are e.g.

- degreasing and cleaning in the metal industry
- painting, lacquering and gluing
- reinforced plastic production
- printing industries
- repair and service operations

Although personal protection is, in principle, the last line of defense it is in many cases the only practical means to protect the worker. Therefore, enough attention should be paid to the proper selection and use of respirators and protective gloves in the handling of solvents. Gloves which allow the penetration of the solvent to be handled or a respirator with unsuitable or saturated filters do not provide protection, indeed, the opposite might be the case, and they only give a false sense of security.

RESPIRATORS

To provide adequate respiratory protection the respirator must be effective enough and it must be properly used. When selecting and using respirators, the following general aspects should be considered:

- identification and evaluation of the hazards concerned
- evaluation of the requirements of the task, such as movements, physical exertion etc.
- selection of sufficient respirators
- instruction and training of workers
- maintenance of respirators (cleaning, inspection, repair and storage)
- medical surveillance

The respirator should not excessively hamper the wearer at work and the additional physiological load on breathing imposed by the respirator should be as low as possible.

Filtering devices with organic vapour filters or combined vapour and particle filters are most commonly used when handling solvents. Other possible types are the compressed air line breathing apparatus and the fresh air hose breathing apparatus.

The facepiece of a filtering device is usually a half mask. Full face mask is used when the ambient concentrations are high or when also the face must be protected against splashes. Filtering devices are relatively small, inexpensive, easy to maintain and they do not restrict the wearer to move. Disadvantages of these respirators are a rather low protection efficiency due to the negative pressure inside the mask during inhalation, a relatively high breathing resistance and the need for frequent replacement of filters. Power-assisted filtering respirators which provide better protection efficiency and a lower breathing resistance have been used against dusts for some time. More recently some manufacturers have introduced also gas and vapour filters for powered respirators. Until now these filters have proved to be expensive particularly when their relatively short service lives are taken into account.

Air line breathing apparatuses are used with a half mask, a full face mask or a hood, helmet or blouse. Compressed air is supplied to the facepiece as a continuous flow of air or through a demand valve or a pressure demand valve. The latter type provides a slight positive pressure inside the mask at all times. The main advantages of the air-line apparatus are a relatively low breathing resistance and a better protection efficiency compared to filtering respirators. In addition, the same equipment provides protection against all contaminants and even oxygen deficiency. On the other hand the air supply hose restricts the mobility of the wearer.

Efficiency of respirators, protection factors

The overall efficiency of a respirator primarily depends on different sources of leakage, such as a face seal leakage, leakage through filters and leakage through valves and fittings (mainly through the exhalation valve).

The overall protection afforded by a respirator may be defined in terms of its protection factor.

$$\text{protection factor} = \frac{\text{concentration in ambient air}}{\text{concentration inside the facepiece}}$$

TABLE 1. Examples of Respirator Protection Factors

Respirator	Protection factor	
	USA	CEN
half mask with gas and vapour filter	10	20
half mask with dust filter	10	9
full face mask with gas and vapour filter	50	2000
full face mask with high efficiency particle filter	50	1000
hose mask without blower, full face mask	50	2000
air line, continuous flow, half mask	1000	1000
air line, continuous flow, full face mask	2000	2000
air line, demand type, full face mask	50	2000
air line, pressure demand type, full face mask	2000	2000
air line, continuous flow, hood, helmet or suit	2000	1000

Protection factors given for some respirators in the

United States (recommended protection factors) and in a proposal for a European norm (nominal protection factors) are listed in table 1 (Hyatt, 1976; CEN, 1984). Nominal protection factors (CEN) are derived from the maximum permitted leakage of the whole equipment (or the sum of the inward leakages for a multicomponent device) stated in different proposed standards. Recommended protection factors used in USA are, on the contrary, based on existing quantitative man-test data. They differ from nominal protection factors especially for respirators with negative pressure inside the mask during inhalation. These protection factors are only guides for the selection of a suitable respirator for a certain task. Individual factors affecting the fit on the face (shape and size of the face, beard, spectacles) should always be considered.

Many field studies have shown lower protection factors than those recommended in table 1. An average protection factor of 3 has been reported for a half mask with a vapour filter in paint spray operations instead of the 10 recommended (Toney, 1976). In polyurethane spray painting the protection factors against airborne isocyanates ranged from 1.4 to 100 when a half mask with a vapour filter was used and from 4 to 500 when the same facepiece with a combined filter was used (Rosenberg and Tuomi, 1984).

To ensure a proper protection the facepiece should be individually fitted. The workers should also be instructed to test the face fit every time when the respirator is worn. Beard and hair between the face and the edge of the facepiece cause a remarkable increase of the face seal leakage (Hyatt et al., 1973; Jönsson, 1980). A marked decrease in the performance of some masks has been observed already when tested on subjects 8 hours after the shaving (Jönsson, 1980).

Service lives of respirator filters

Activated carbons used in organic vapour filters have a limited adsorption capacity. After initial breakthrough the effluent concentration increases quite rapidly and the filter should be replaced.

The service life of a respirator filter is defined as the time during which the breakthrough concentration has attained a certain fixed value. Threshold Limit Values or corresponding hygienic standards are commonly used for service life determination (Balieu, 1976; Miller and Reist, 1977). Also fixed lower breakthrough concentrations or certain percentages of the effluent concentration have been used (Freedman et al., 1977; Jaakkola, 1978; Nelson and Harder, 1974, 1976).

The most important determinants of the service life of activated carbon filled filters are filter characteristics (properties and amount of carbon, dimensions of carbon bed), conditions of filter usage (type and concentration of the solvent, relative humidity) and storing conditions. Manufacturers use different types and amounts of carbon. Therefore the service lives of different filters may vary considerably even under the same conditions.

In official certification tests organic vapour respirator filters are classified in most European countries into three classes according to their capacity and breakthrough times. Usually this filter class is the only piece of information regarding the performance available to the user. To estimate the service life on the basis of the filter class only is an impossible task. Usually the old rule of thumb to change the filter when the odour of the contaminant vapour becomes detectable, is used. This method is not safe for odourless vapours or vapours with weak odour (Reist and Rex, 1977). Furthermore, the olfactory fatigue and many other individual factors make this method unsafe. In practice the filters are used all too long, sometimes even for many months.

Many studies have revealed that service lives of organic vapour respirator filters are relatively short. Nelson and Harder determined the service lives of a filter containing 55 g carbon for 121 solvents (concentration 1000 cm^3/m^3, flowrate 53 l/min). The 1% breakthrough times varied from 0.05 minutes for methyl chloride to 143 minutes for 1-nitropropane (Nelson and Harder, 1974). Methanol, vinyl chloride, ethyl chloride, dichloromethane and methylamine had breakthrough times shorter than 15 minutes. Service lives were longer than 90 minutes for e.g. toluene, perchloroethylene, xylene and butanol. Within each class the most volatile solvents had the shortest breakthrough times. The same authors have also shown for 9 solvents that the ambient concentration (C) and the time to reach any fixed breakthrough percent (t_b) are interdependent according to the expression

$$t_b = a \cdot C^b,$$

where a and b are constants that depend on experimental conditions (Nelson and Harder, 1976). According to their results the average value for b at 10% breakthrough was $- 0.67 \pm 0.17$. The service life is inversely proportional to flowrate and is shortened at relative humidities greater than 50%.

With twelve different filters the 1% breakthrough times for toluene (1000 cm^3/m^3, 30 l/min) varied from 44 minutes to 289 minutes (Jaakkola, 1980). The 1% breakthrough times of one filter type at toluene concentrations of 1000, 500, and 200 cm^3/m^3 were 229, 436 and 855 minutes, respectively. Four of these filters were also tested for styrene. Corresponding 1% breakthrough times (1000 cm^3/m^3, 30 l/min) varied from 233 minutes to 387 minutes being 30-80% longer than for toluene. Remarkable desorption of both toluene and styrene was found when filters saturated up to 50 and 100% breakthrough were tested with clean air. Intermittent testing procedure (periods of 0.5 and 2 hours daily) simulating the normal use of a filter did not markedly change the service lives obtained in a continuous testing set-up.

Although breakthrough times are longer in lower concentrations, it can be concluded that service lives of organic vapour filters are relatively short. In typical tasks involving solvent handling the service lives are 5-20 hours depending on the type and concentration of the solvent. For the most volatile solvents the service lives are remarkably shorter. Therefore, filtering respirators should not be used against methanol, vinyl chloride, ethyl chloride, dichloromethane, and methylamine vapours. At high relative humidity the adsorption of water vapour to activated carbon shortens the effective service life of the filter.

The Labour Protection Boards both in Finland and Sweden have proposed the following simple formula to be used when estimating service lives:

$$t = \frac{1\,000\,000 \cdot G}{V \cdot C}$$

where C = concentration of the contaminant, mg/m^3
V = breathing rate, l/min (depends on heaviness of work)
G = adsorption capacity of the filter, g

Although this equation overlooks the differences both in the properties of the filters and in the adsorption characteristics of different solvent vapours, it can be used to get a rough estimate of the service life when experimental values are not available. When this formula was introduced in Finland and in Sweden the adsorption capacities given for organic vapour filters were 18 g for class I filter and 35 g for class II filters. Calculated service lives based on these adsorption capacities were in many cases longer than the empirical ones (Jaakkola et al., 1984). Recently recommended mew adsorption capacities are considerably smaller,

$$G = 4 \text{ g for class I filter}$$
$$G = 19 \text{ g for class II filter.}$$

These values have their basis in the certification requirements (minimum capacity measured with carbon tetrachloride) of the proposed European Standard for gas filters and combined filters (CEN, 1981). When these smaller adsorption capacity-values are used in calculations the risk of obtaining excessively long service lives is much smaller.

One further point to note is that respirators with only gas or vapour filters provide little or no protection against aerosols. Filtering efficiency of 4 different organic vapour filters against aerosols with particle diameters of 0.3 and 3 µm ranged from 7.5 to 25.9% and from 71.6 to 98.0%, respectively (Jaakkola et al., 1984).

Conclusions on respiratory protection

1. The facepieces should be individually fitted and different facepieces should be available to ensure a properly fitting mask for every worker supposed to wear a respirator. A good sealing between the face and facepiece is especially important with normal filtering respirators because the negative pressure inside the mask during inhalation makes them sensitive to inward leakages.

2. Organic vapour filters should be replaced regularly. The fact that filters' service lives are relatively short must be kept in mind.

3. Filtering respirators should not be used against the most volatile solvents like methyl chloride, vinyl chloride and dichloromethane (methylene chloride).

4. A combined filter is necessary when the air contains both vapours and aerosols. This is the case e.g. in spray painting.

5. Filtering respirators do not offer adequate protection for persons with a full beard or sideburns.

6. In continuous work at stationary work sites air-line apparatuses are recommended because they provide better protection and cause smaller physiological strain than the filtering respirators. In addition, no need exists for frequent filter replacement.

7. Usually the concentrations of oil mist and hydrocarbon vapours in the industrial compressed air networks are so high that the air, if used for breathing air, should be purified with a filtering device that contains an effective particle filter and an activated carbon filter.

8. Fresh air hose breathing apparatuses with a blower are recommended at temporary work sites where the concentrations of solvent vapours are high and where compressed air is not available.

9. Respirators should be cleaned regularly. Special attention should be paid to the inhalation valve which, if dirty or damaged, might lead to remarkable inward leakages.

10. When not in use, respirators should be stored in clean, uncontaminated boxes or lockers.

PROTECTIVE GLOVES

In direct skin contact solvents defat the skin and may cause contact eczema. Some solvents in liquid form (e.g. toluene, xylene, butanol, methanol, dichloromethane and styrene) also penetrate through the skin so rapidly that the total exposure may significantly increase (Engström et al., 1977). If skin is diseased or damaged in some way, solvents will penetrate more easily (Riihimäki et al., 1978). The main contact areas are the hands and forearms which are exposed during many routine solvent handling operations. If technical control measures and appropriate work practices prove to be an inadequate safeguard, proper gloves made of natural or synthetic rubber or plastic materials should be used to minimize the contact.

The first thing to consider when selecting gloves for solvent handling is the chemical protection afforded by the glove. In general, the permeability of the glove material to the solvent in question should be as low as possible. Gloves fully impregnated with solvents or those allowing the permeation of solvents after a short period of use can be more harmful than working without gloves. In the former case the solvent comes into direct and continuous contact with the skin. Furthermore, the high humidity and raised temperature of the skin inside the gloves will likely increase the permeability of the skin manyfold (Bird, 1981).

In addition to permeability, other important factors to consider are the possible allergenic additives of the glove material, the mechanical properties of the material and the suitability of the glove both to the hand and to the work. Rubber gloves (both natural and synthetic) in particular contain allergens in the form of accelerators and antioxidants. Finnish statistics of occupational skin disease indicate that about 60% of the annually registered new rubber eczemas are due to rubber gloves (Estlander et al., 1984). Rubber polymers in itself are seldom contact allergens but can sometimes induce contact urticaria. Plastic materials cause very seldom contact allergic eczema.

Performance of gloves against solvents

The gloves are usually selected on the basis of information found in manufacturers' catalogues or handbooks which reveal only general qualitative ratings on the properties of the gloves or materials. These ratings do not provide information on the permeation rates and breakthrough times for solvents. Therefore, the selection of gloves for a certain task is all too often a trial and error process which can imply considerable problems of skin protection. The best results are achieved if gloves can be selected on the basis of actual permeability data.

Breakthrough times and permeation rates of many liquid solvents for the most commonly used rubber and plastic gloves have been determined in several studies. Many of these have been done with circular sections cut from the flat surfaces of the gloves.

In a comprehensive survey Nelson and his coworkers tested the penetration of 29 common solvents through 28 different types of commercially available protective gloves (Nelson et al., 1981). None of the gloves tested offered a satisfactory protection against all the solvents. Breakthrough times for polyethylene gloves (disposable type), PVC gloves and natural rubber latex gloves were quite short and exceeded 15 minutes in only some cases. Neoprene and nitrile (NBR) gloves had, in general, longer breakthrough times. However, methylene chloride, methyl iodide, tetrahydrofurane, chloroform, trichloroethylene and ethylene dichloride penetrated even these gloves rather quickly the longest breakthrough times being 13, 17, 20, 21.6, 26 and 28 minutes, respectively. At least one glove type with a breakthrough time longer than 30 minutes was found regarding the others of the 29 tested solvents. NBR gloves were more resistant against aromatics and the neoprene gloves were resistant against acetone. Marked differences were observed in the permeability of gloves made of the same material by different manufacturers.

Our own permeability studies on toluene with 17 different types of gloves and on trimethylpentane with 6 types of gloves yielded similar results. PVC gloves allowed a breakthrough relatively rapidly and stiffened after the testing with both solvents. Toluene permeated a disposable PVC glove almost immediately. Nelson observed somewhat longer breakthrough times with two types of disposable gloves made of polyethylene (Nelson et al., 1981).

In practice protective gloves are not in a continuous contact with liquid solvent. The duration of an adequate protection is thus longer than the breakthrough times reported above and it depends on the total solvent contact time. However, it should be considered that the permeation of solvent splashes resembles the permeation of liquid solvent (Sansone et al., 1981).

Besides empirical screening tests, solubility parameters have been used to predict the permeability characteristics of protective clothing. This theory which has been succesfully used in the formulation of coatings has also given promising results in the selection of the right material for protective clothing (Hansen, 1984; Henriksen, 1984).

Barrier creams are sometimes recommended as substitutes for protective gloves. However, the permeability of many barrier creams to organic solvents is high (Steen, 1984). Furthermore, the rather thin film of barrier cream applied on hands becomes easily worn away during the work. Hence, barrier creams constitute no alternatives to protective gloves.

Conclusions on skin protection

1. Selection of protective gloves against a certain solvent should, if possible, be based on relevant permeability data. Differences in the performance of gloves made of the same material by different manufacturers should also be considered.

2. All commercially available gloves are permeated by most of the commonly used solvents after a certain time. Gloves coming into continuous or repeated shorter contacts with the liquid solvent should be changed regularly.

3. PVC and natural rubber are not recommended because of their relatively high permeability to almost all common solvents. The best materials in many cases are either neoprene or nitrile rubber or Viton elastomer. Polyvinyl alcohol would give the best chemical protection but it is not practical because of the poor resistance to water (Williams, 1979).

4. Allergic or irritant eczemas may ensue from the use of rubber gloves, requiring proper attention.

5. If disposable PVC or polyethylene gloves are worn due to the requirements of the job, they should be changed after every direct contact with the solvent (also splashes).

6. Textile gloves should be used under protective gloves that do not have an interior liner.

7. Before use new gloves should be inspected carefully for pin holes, cracks, thin spots and other possible damages of the polymer film.

8. Barrier creams which are sometimes recommended as substitutes for protective gloves do not offer adequate protection against organic solvents.

REFERENCES

Balieu E (1983). Respirator filters in protection against low-boiling compounds. J Int Soc Resp Prot 1:125-138.

Bird MG (1981). Industrial solvents: Some factors affecting their passage through the skin. Ann Occup Hyg 24:235-244.

CEN (1981). Respiratory protective equipment: Gas filters and combined filters, requirements, testing, marking. Document prEN 141, edition 1.

CEN (1984). Respiratory protective equipment: Guidelines for selection of respiratory protective equipment. Document CEN/TC 79/SG1 90.

Engström K, Husman K, Riihimäki V (1977). Percutaneous absorption of m-xylene in man. Int Arch Occup Environ Hlth 39:181-189.

Estlander T, Jolanki R, Kanerva L (1984). Disadvantages of rubber and plastic gloves. Scandinavian symposium on protective clothing against chemicals. Copenhagen, Denmark 26-28 Nov 1984, conference abstracts.

Freedman RW, Ferber BJ, Hartstein AM (1973). Service lives of respirator cartridges versus several classes of organic vapours. Am Ind Hyg Ass J 34:55-60.

Hansen CM (1984). Solubility parameters. Polymer/solvent interactions, estimation of parameters. Scandinavian symposium on protective clothing against chemicals. Copenhagen, Denmark 26-28 Nov 1984, conference abstracts.

Henriksen HR (1984). Solubility parameters. A starting point in prediction of chemical resistance. Scandinavian symposium on protective clothing against chemicals. Copenhagen, Denmark 26-28 Nov 1984, conference abstracts.

Hyatt EC (1976). Respirator protection factors. Report LA-6084-MS, Los Alamos, New Mexico: Los Alamos Scientific Laboratory.

Hyatt EC, Pritchard JA, Richards CP, Geoffrion LA (1973). Effect of facial hair on respirator performance. Am Ind Hyg Ass J 34:135-142.

Jaakkola J (1980). Service lives of different organic vapour respirator filters. Proceedings of the International symposium on air pollution abatement by filtration and respiratory protection. Copenhagen, Denmark, 4-6 Nov 1980, collected papers 1.

Jaakkola J, Tuomi T, Tossavainen A, Korhonen E (1984). Hengityksensuojainten suojausteho (The efficiency of respirators), report no 207. Helsinki, Finland: Institute of Occupational Health (in Finnish with an English summary).

Jönsson PG (1980). Effect of beard, beard growth and age wrinkles on respirator performance. FOA rapport A 40034-C2 (A2,B2), Umeå, Sweden (in swedish).

Miller GC, Reist PC (1977). Respirator cartridge service lives for exposure to vinyl chloride. Am Ind Hyg Ass J 38: 498-502.

Nelson GO, Harder CA (1974). Respirator cartridge efficiency studies. V Effect of solvent vapour. Am Ind Hyg Ass J 35: 391-410.

Nelson GO, Harder CA (1976). Respirator cartridge efficiency studies. VI Effect of concentration. Am Ind Hyg Ass J 37: 205-216.

Nelson GO, Lum BY, Carlson GJ, Wong CM, Johnson JS (1981). Glove permeation by organic solvents. Am Ind Hyg Ass J 42: 217-225.

Reist PC, Rex F (1977). Odor detection and respirator cartridge replacement. Am Ind Hyg Ass J 38:563-566.

Riihimäki V, Pfäffli P (1978). Percutaneous absorption of solvent vapours in man. Scand J Work Environ Hlth 4:73-85.

Rosenberg C, Tuomi T (1984). Airborne isocyanates in polyurethane spray painting: Determination and respirator efficiency. Am Ind Hyg Ass J 45:117-121.

Sansone EB, Leonard AJ (1981). Resistance of protective clothing materials to permeation by solvent "splash". Environ Res 26:340-346.

Steen L (1984). Barrier creams - a comparative study of in vitro barrier effect. Scandinavian symposium on protective clothing against chemicals. Copenhagen, Denmark 26-28 Nov 1984, conference abstracts.

Toney CR, Barnhart WL (1976). Performance evaluation of respirators used in spray painting operations. HEW Publication no (NIOSH) 76-177, Cincinnati.

Williams JR (1979). Permeation of glove materials by physiologically harmful chemicals. Am Ind Hyg Ass J 40: 877-882.

Index